Implementing CIM

IMPLEMENTING CIM
COMPUTER INTEGRATED MANUFACTURING

Anna Kochan
and
Derek Cowan

IFS (Publications) Ltd, UK

Springer-Verlag
Berlin · Heidelberg · New York · Tokyo

Anna Kochan
IFS (Publications) Ltd
35-39 High Street
Kempston
Bedford MK42 7BT
England

Derek Cowan
The Plessey Company plc
(Plessey Research and Technology)
Roke Manor
Romsey
Hampshire SO5 0ZN
England

British Library Cataloguing in Publication Data

Kochan, Anna
    Implementing CIM.
    1. Computer integrated manufacturing systems
    I. Title   II. Cowan, Derek
    670.42'7   TS155.6

ISBN 0-948507-20-9 IFS (Publications) Ltd
ISBN 3-540-16352-2 Springer-Verlag Berlin Heidelberg New York Tokyo·
ISBN 0-387-16352-2 Springer-Verlag New York Heidelberg Berlin Tokyo

Phototypeset by Wagstaffs Typeshuttle, Henlow, Bedfordshire
Printed by Bartham Press Ltd, Luton

CONTENTS

Anna Kochan, graduated from Sussex University with a degree in mechanical engineering in 1976. Having trained as a technical journalist on The Engineer and Metalworking Production, magazines published by Morgan Grampian, she joined the publications department of the Institution of Production Engineers and became Editor of Numerical Engineering and Deputy Editor of the Production Engineer. She now works for IFS Publications as Editor of The FMS Magazine, FMS Update and Assembly Automation.

Derek Cowan is the Group Manufacturing Technology Executive for The Plessey Company plc, England. An engineering graduate of Liverpool University, he also studied at the McGill University, Montreal, and the Northeastern University, Boston. His 30 years experience, as an engineer and manager, in the semiconductor and electronics industries in North America and the UK include positions as a Senior Product Engineer, Fairchild Semiconductors; Product Manager for integrated circuits, Elliott Microelectronics; Manufacturing Manager, Hewlett-Packard; General Manager, Plessey Secure Digital Systems; and Site Operations and Business Systems Executive for the Plessey Defence Systems Division.

Preface

This book has been written to fill a gap in the market. While there are many textbooks which discuss in great detail the technologies of CIM, there is none which presents the total concept in a quick, easy-to-read style acceptable to managers.

Managers have to know what CIM is and how it affects the whole manufacturing process but they do not need to be burdened with the minute technical detail – this can be left to the experts in engineering, information technology, personnel and finance. People caught up in a CIM implementation, however, would also find this book of interest. For instance, an engineer may be an expert in one or other of the many CIM technologies, but because of the integrated nature of CIM, he must have a general appreciation of the others. Only with an understanding of the total concept, can the engineer help achieve smooth integration.

This book, therefore, aims to provide a complete overview of the concept of CIM in a clear and comprehensible style, without getting immersed in technical detail. It has been written by a professional author, Anna Kochan, and a professional manager, Derek Cowan, whose responsibility for CIM strategy for The Plessey Company has contributed a real-life emphasis to the book.

Chapter 1 is an introduction to CIM: the technology and its implementation. Chapter 2 puts CIM into the context of the development of manufacturing, discussing how the advent of computer control has affected the organisation of manufacturing. The subsequent seven chapters each address different CIM enabling technologies: computer aided design, computer aided manufacture, computer aided test, manufacturing planning and control, process technologies, robotics, and automated material handling, emphasising how each fits into the overall CIM strategy. Chapter 10 focuses on the backbone of CIM implementations, the communications. This is a fast-changing field, particularly concerning standards. Chapter 11 brings the whole subject together, and examines the management support aids to developing a CIM strategy. Finally, a bibliography and glossary of terms complete the book.

By reading this book, you will not become an expert on CIM, but you will develop an understanding of what CIM is and become familiar with the major issues involved.

Finally, the authors would like to thank Dickie Davies of Plessey (now retired) for his help with Chapter 7, Peter Tullett of Plessey for his input on Chapter 5, and most of all David Liddell, the Chairman of the Plessey Methods Panel (of which the IMP – Integrated Manufacturing Project managed by Derek Cowan was a part) whose initiative was responsible for the book's conception.

1 INTRODUCTION TO CIM

Introducing new technology requires a structured approach

The manufacturing environment is changing dramatically as the microprocessor finds applications in every facet of the manufacturing organisation. But, if allowed to invade on a random basis, the microprocessor is unlikely to provide a company with optimum operating conditions.

Many manufacturing concerns already use computer equipment to considerable advantage, particularly in design and drafting, production control, machine control and order processing. In all these applications, the computer has been employed merely to perform a job previously done manually, in the same way but more efficiently. Today many manufacturers are beginning to realise that it is vital to have a structured approach to the

A strategy for implementing CIM must be formulated NOW

introduction of this technology. One reason for this is that, using computers, it is likely that a completely different way of operating may be more efficient. The other reason is the tremendous advantage to be gained by enabling computers in different sectors of the company to communicate and exchange information. Companies that have already formed this conclusion are not finding it easy to put into practice because of the difficulties of linking different vendors' products. It is, therefore, imperative to formulate a strategy for Computer Integrated Manufacturing (CIM) at the earliest possible stage.

There is no standard formula for CIM – each company must define its own

There is no standard formula as to how a company should adopt computers to create a CIM plant. All companies operate in different ways and in different market places, and each one must devise its own strategy for success.

The overall objectives of every company include maximising profit. This is achieved by maximum utilisation of assets and optimum customer satisfaction, i.e. on-time delivery of good quality products at a competitive price. The ultimate aim of CIM is to produce the correct number of parts of acceptable quality at the right time. The ranking of the individual priorities will depend on the company's own macro strategy.

CIM includes everything – from tendering to post-delivery service

As one of the latest manufacturing buzz words, there is often some confusion, and even commercial abuse, over the use of the term computer integrated manufacturing. Firstly, it is important, when talking about manufacturing, to include not only the production activities but also marketing, sales, engineering, materials, finance and personnel – in fact every activity within the organisation, from tendering to post-delivery service (Fig. 1).

Companies that have decided on a future in manufacturing are under great pressure to introduce new technology. Many are beginning to look towards the 'factory of the future' – a philosophy encompassing the integration of projected future technologies to manufacture products in a peopleless, paperless environment.

CIM is a strategy, incorporating computers, for linking existing technology and people to optimise business activity

CIM can be thought of as the initial strategy for achieving the 'factory of the future'; it is defined as a

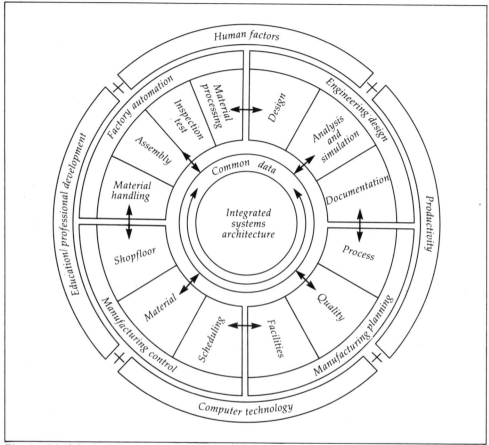

Fig. 1   Model of a CIM system (courtesy of SME)

strategy for linking existing technology and people to optimise business activity.

CIM STRATEGY

Analyse needs and objectives before forming a plan

A company's business ambitions will determine its individual CIM strategy. A company must, therefore, first analyse its business and determine a priority of business objectives before it can start to draw up a plan for CIM. A good quality strategy will have the robustness to respond to any change of priorities.

Some examples of business objectives are:

- Fast response to market demands
- Better product quality
- Reduced cost
- Enhanced performance
- Better asset utilisation
- Shorter development lead times
- Minimum work-in-progress
- Flexibility

It may be useful to split these objectives into management and operations sections.

Having assigned priorities to the business objectives, a company can begin to form a strategy to achieve them. The strategy will concern every aspect of the manufacturing organisation, and will start with the 'as-is' analysis.

The impact of CIM touches every part of the company's activities, and this means that all personnel in the company, whether they are engineers or accountants, fork-lift truck drivers or managers, need to be aware of the changes going on around them. The unmanned factory will never be realised completely. It is more likely to be human barriers rather than technological barriers which will delay full implementation of CIM.

Top management commitment is vital for success

It is mandatory to have complete top-management commitment to the introduction of CIM if it is to be successful.

ENABLING TECHNOLOGIES

The enabling technologies involved in CIM are essentially the following (Fig. 2):

- Computer aided design (CAD)
- Computer aided manufacture (CAM)
- Computer aided test (CAT)
- Manufacturing planning and control
- Process technologies
- Robotics
- Automated materials handling

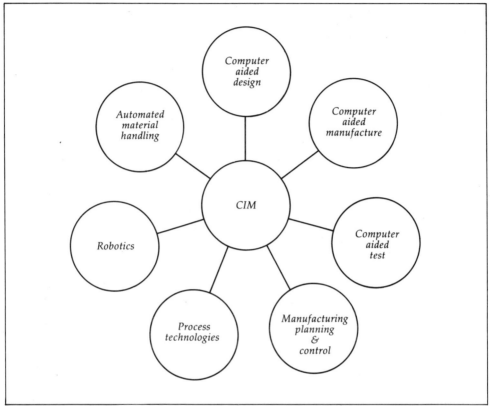

Fig. 2   The enabling technologies of CIM

with communications linking the transfer of data. (A chapter of the book is devoted to each of these topics.) Fig. 3 shows how the enabling technologies relate to the objectives of CIM.

CIM is unlikely to be a turnkey operation

Unless there is a greenfield site, the implementation of the CIM strategy is unlikely to be a turnkey operation due to the enormity of the task. It is most likely to be incremental, transforming the facility into a 'factory of the future' on an evolutionary rather than a revolutionary basis. In some areas, completely new technology will have to be introduced, but in most others, existing 'islands' of automation will need to be linked. This is currently precluded by the inability to transfer the necessary data.

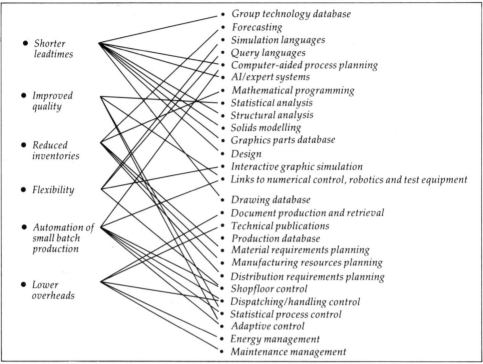

Fig. 3 Relationship between CIM technologies and objectives

As the individual activities are modernised progressively, data management and network communications capabilities must be introduced to facilitate data transfer between the various production, engineering and business activities. It is vital in any CIM implementation that a communications network exists giving all areas of the organisation access to the same data. A functional diagram of the manufacturing enterprise is shown in Fig. 4.

A communications network must give everyone access to the same database

COMMUNICATIONS

The communications network is the most crucial single element of a CIM implementation – it is also the one which causes the most problems. Generally there are a minimum of four levels in a CIM communications hierarchy (Fig. 5):

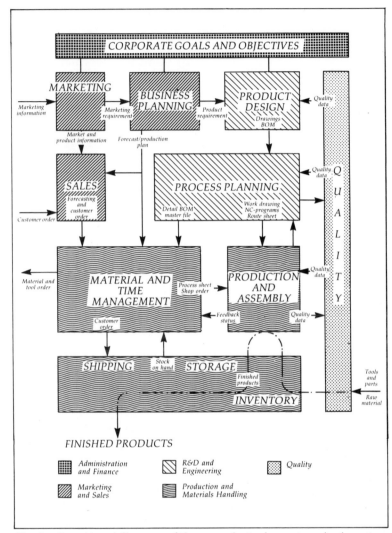

Fig. 4   Functional diagram of the manufacturing enterprise (courtesy of Digital Equipment Corp.)

- Business mainframe
- Factory host
- Shopfloor controller
- Process controllers (robot, AGV, machine tool controls, etc.)

*Plan the final
phase before
purchasing the
first*

The problems of communications essentially relate to the mix of vendor products, both hardware and software, which any company is bound to need in its application of CIM. Different vendors' products are often incompatible to the extent that communication between them is impossible. Therefore, when planning a CIM strategy, companies must take care to specify the systems down to the last detail to ensure that full integration is possible. This is a particular warning to companies intending to implement in a phased mode – they must plan the final phase in full before purchasing the first.

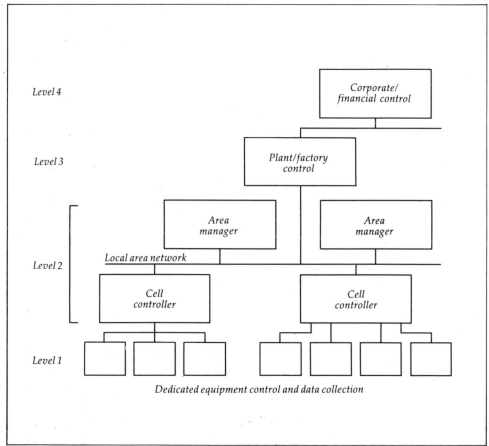

Fig. 5   The four levels of the control hierarchy (courtesy of Hewlett-Packard)

DATA PROCESSING

Don't get carried away with creating data

With a communications network established, data can be transmitted around the company; it can be processed and used to create more data for control or decision purposes. It is important not to get carried away with collecting and processing data, as the real point of the exercise is to create profit. Companies must be disciplined to define the functional relationships and data relationships clearly, so that the sources and users of all data can be traced. (A suitable methodology for doing this, called 'IDEF', is explained in Chapter Eleven.) And if data is accumulating for no useful purpose, then it must be eliminated; therefore, a data management and control system is needed.

It is now possible to collect or extract selected data from different functional areas, such as engineering, manufacturing or finance, to process that data using office automation techniques, and finally to transmit a report by electronic mail. An individual can therefore control his own requirements for data collection, analysis, presentation and communication, and the need for the classical electronic data processing (EDP) organisation no longer exists.

On the topic of data protection, all computer equipment suppliers are beginning to incorporate sufficient security in their products to limit who can change or who can use which data. This is particularly important for those involved in Ministry of Defence contracts who need to keep classified data under tight security, giving access only to those who need it and in some cases no access at all.

VENDOR RELATIONSHIPS

Cooperation of vendors and workforce is vital

The manufacturing system which results when CIM is implemented is very finely tuned. To maintain profitable conditions, the system must be able to retain its balance whatever unpredictable events should occur. It is likely that a system will contain some redundancy to account for machine breakdown or market changes. However, there are two bodies of people whose cooperation the system is totally reliant on: its vendors and its workforce. No amount of good planning can produce a manufacturing system able to function without materials or manpower.

Few companies will be able to establish a relationship with their vendors such that they will deliver the relevant goods during a set four-hour period each week, as one US company has organised. However, there are many ways in which a vendor's goodwill can be obtained; the end objective is to have a good vendor-user coexistence relationship.

Good vendor-user coexistence is the objective

The vendor needs to know with some confidence what a company expects over a period of time, so that a schedule can be planned accordingly. The company should outline its expected monthly requirements from each vendor over an 18-month period. It should agree a variation magnitude in the contract and as each delivery date moves closer, tighten the tolerance. Also, should orders fall down just before delivery, the company must accept the orders rather than cancelling the day before. Another aspect of good vendor relationships is prompt payment – the vendor also has to think about his cash flow!

One requirement that companies should impose on their vendors is 100% quality of incoming goods. Material arriving at the factory should be able to go straight on the line without prior inspection.

EMPLOYEE RELATIONSHIPS

Working practices must change to reflect new technology

CIM presents many changes in working practices to a company's workforce. Job demarcation will become less specific because of the integration of the different functions. All employees will have to become accustomed to receiving instructions in graphical form and to report to computer terminals.

For these changes to take place smoothly, the workforce must be informed of the intention to purchase any new technology at the very earliest opportunity, i.e. as soon as the decision to purchase has been made. Training in any new skills required must be offered to all those affected by the purchase and anyone else interested in acquiring them. The workforce must feel involved in the developments going on around them; the change should proceed in an atmosphere of excitement!

A factory operating CIM principles infers that its personnel have a broader spectrum of knowledge than previously required. Companies are now starting to pay employees to acquire knowledge and experience for jobs which they may only be called upon to undertake when the normal operator is absent.

FINANCE

Few CIM investments satisfy classical return-on-investment criteria

The implementation of CIM for any one company is effected at considerable expense – which is not easy to justify using traditional financial yardsticks. In fact, few, if any, of the current CIM projects would satisfy classical 'return-on-investment' criteria. It is no good insisting that a capital investment relating to CIM has a better than two-year payback period. In the final analysis, a CIM project with a seven-year payback period could give a better solution. Consideration of the intangible factors which are not normally taken into account have now been incorporated into financial evaluation packages, such as IVAN (developed at the University of Manchester Institute of Science and Technology (UMIST)). (This also incorporates a 'what-if' capability.)

Investment in CIM is essential to achieve the company's long-term objectives

The investment in CIM, which generally runs into several million pounds, must be thought of as essential to achieve the company's long-term objectives. Since the implementation will be evolutionary, a long-term commitment to the investment must be made, and a company must include CIM funding in its forward planning. This can be complicated, as many of the enabling technologies to be incorporated in the future may not be currently available products and systems.

Many companies have set up 'islands' of automation – cells of robots for paint spraying, production control systems, automated warehouses, computer-aided design systems, flexible machining systems, and so on. Investment in these areas has been most significant in the automotive industry and in the aerospace and defence industries. None has yet achieved the integration of the entire manufacturing process although moves to accomplish this are in progress.

2 CIM IN CONTEXT

*Evolution in
manufacturing
technology has
brought about CIM*

But how has CIM evolved? Before computers were used to control manufacturing processes and to process data, machines were operated primarily by people according to verbal or written instructions. The operator turned the handles, pulled the levers and pushed the buttons, using skills often acquired during a long apprenticeship. The operator was a craftsman; he or she knew what allowances had to be made for material and machinery.

The operator was responsible also for any data collection, which was often presented in a terminology only understood by the foreman and other craftsmen. The environment relied upon the foreman and the craftsmen, who were almost totally isolated from management.

PHASE 1
The manual press-brake
Economic batch quantities are handled
The gauge is set manually
One bend per pass is formed
All operator information is on paper

PHASE 2
The NC press-brake
Lot sizes of 1 are handled
The gauge moves automatically by means of a stepper motor under computer
control
All bends are formed in one pass
A terminal provides the operator with all necessary data

Fig. 6 Evolutionary operation of a press-brake typifying the development of CIM

At the turn of the century, Taylor formulated the principles of manufacturing control, based on the premise that the operators were uneducated, ill-disciplined and unmotivated. He proposed that workers had to be directed and rigorously monitored and controlled; the 'what' and 'when' to do a job had to be instructed, the 'how' having been learnt during the apprenticeship.

Written and verbal communication is being eliminated

Much of industry still operates on these lines, but companies are gradually changing. Written and verbal data communication is being eliminated, the human role is changing drastically, and more interesting employment in more pleasant environments is emerging.

A simple example which illustrates the fundamental changes taking place in industry is the evolutionary operation of a press brake for the folding of sheet metal (Fig. 6).

MANUAL OPERATION

In its original form, the press brake required an operator to be in attendance at all times. Each day a list of jobs, in the form of a sheaf of papers, would be given to the operator. Each job would require a shop order produced by the production control department, stating the job number and quantity required, as well as a process sheet from the engineering department describing the configuration of the parts. This would be accompanied by a drawing relating to each job.

Equipped with his instructions, the operator would proceed with setting up the press brake for the first fold. This would be done by manually setting a back gauge finger at the required distance behind the press-brake tooling such that when the material was fed up against the gauge and the press-brake ram operated the metal would be folded in the required configuration. This distance had to allow for the thickness of the sheet metal, material temper and tooling variation.

Having set the machine, the operator would complete the first fold on each of the parts in the batch one-by-one. As each part came off the machine, he would pile them up alongside the machine in preparation for the subsequent folds. When the first fold had been made on the entire batch, the operator would set up the machine for the second fold and repeat the process as many times as necessary to complete the job. Thereafter, the operator updated the paperwork to show job duration (often split between set-up and run times) and excess allowances claimed. This paperwork then went with the job on to the next process.

Manual operation meant economic batch quantities

The main disadvantage was that the first fold had to be made on the entire batch before the second, and so on, making scrap production more likely. There are two main reasons for this: when each fold is set up on the machine there is a small trial and error element and each set-up will result in one or two scrap pieces; also, there is a probability that, when the stage of setting up the final fold is reached, a discrepancy would be found in the dimensions, because errors have accumulated owing to variations in the quality of blanks and original material, damage to a part during interim processing, and slippage of the gauge.

Since the set-up time is a significant part of the operation, there was a tendency to introduce economic batch quantity (EBQ) lot sizes, scheduled on an order-point basis. Further disadvantages were thus created by the significant work-in-progress value and the inherent storage and handling problems associated with EBQs. The EBQs would have to make allowances for spares and subsequent operation loss to ensure that sufficient numbers of acceptable parts were produced.

In addition to the disadvantages outlined above, industrial relations difficulties could easily occur, particularly as a result of disputes between the operator and the engineering department over the set-up and run times shown on the process sheet. Long lead times, high inventory, high work-in-progress, poor spares ranging, complex storage/handling, higher scrap, and quality difficulties also made this method less attractive.

NUMERICAL CONTROL

Numerical control improved the economics of small batches

The next evolutionary stage in the operation of the press brake was the application of numerical control (NC), which enabled the position of the back gauge to be dialled in (Fig. 7). The gauge, in this case, was an automatic point-to-point system which bolted onto the press-brake bed. The electronics in the NC monitors the gauge position with a closed-loop feedback technique. It can be given six or more different bends to perform, each of which may be repeated several times for a production job.

The main advantage of this development was that all the folds on one part could be completed in a continuous operation. The operator could check all bends on the first completed part and make an optimum trade-off of dimensions before working on all the other parts in the batch. As well as cutting down on scrap, the system reduced both set-up and handling time and dramatically increased productivity. On runs of up to 200 pieces, productivity was claimed to increase by a factor of three to five, even on parts with only two bends.

Thus, in summary, the benefits of NC are:

- The ability to make a batch of one
- Reduced work-in-progress
- Improved quality
- Reduced set-up time and hence reduced lead-time

Fig. 7   CNC press-brake complete with pendant type control console, pedestal two hand control and foot pedal (courtesy of The Press & Shear Machinery Co. Ltd)

The operator using NC, however, still receives instructions on paper and, if he/she has any previous experience with similar jobs, relies on memory for any information that might be useful for current jobs.

COMPUTER INTEGRATED MANUFACTURING

With the move to computer integrated manufacturing, the operator still stands at the machine but is equipped with a terminal giving him access to all the information and instructions relating to all the aspects of manufacture at that location.

Each operator in the CIM environment is provided with a monitor and keyboard which is linked into the

CIM links the operator into a communications network

communications network which forms the backbone of the integrated factory. This workstation, situated alongside the press-brake, has a direct link with the production control and engineering departments. It also interfaces with the computer control on the press brake.

The production control department is 'master' of the press-brake operation. Via the video monitor, it tells the operator which parts to produce in what quantities and when to make them. It also ensures that the raw material required to perform the job is delivered in the right quantities at the required time. The NC programs are prepared in the engineering department and down-loaded from the engineering database to the press-brake control, via the communications network, in time for their commencement. Any information regarding past experience with the same or similar batches of work are also displayed on the monitor for the operator's information.

The operator in the CIM environment now has a very different job to perform. Having first identified himself/herself to the computer system to gain authorisation to proceed, he/she must tell the production engineering department via the network whether the press-brake is operable (in case of previous maintenance orders). Once this is ascertained, instructions relating to the first job will appear on the monitor. Instructions will be relayed simultaneously to the press-brake control, so all the operator needs do is feed the raw material to the machine, initiate the cycle and remove processed parts. At the end of the batch, any details relating to problems with the machine or with product quality must be entered via the workstation keyboard; and the operator is then ready to start the next job.

Total control of the operation is possible with CIM

The most important benefit of this CIM set-up is that total control of the process is possible; it leaves little room for any error to occur and ensures that information (and not just a narrow area of information) is constantly updated. For example, raw materials inventory will automatically be updated and the production schedule will be informed that the process is complete.

For greater control, bar codes can be applied to the raw material and a bar code reader interfaced with the workstation so that when the operator starts a new job he can check that the raw material supplied tallies with the work order information. In addition, the press-brake microprocessor control system will automatically count the units produced and check that this is consistent with the job order.

The method of working described above leaves no element of chance in the manufacturing process. With the high level of control it is possible to run the operation almost at the limits of its capabilities. The principles described here must be applied to every part of the manufacturing plant to achieve the total CIM situation.

CIM eliminates all elements of chance

3 COMPUTER AIDED DESIGN

Computer aided design (CAD) technology is often thought of in terms of the representations of a product (or part of a structure) that can be displayed in multiple colours on a screen and modified instantaneously as product changes are introduced.

The essence of CAD is the product database

This view of CAD has some justification, but it does not emphasise the most important features particularly when considering CAD in the CIM environment. The essence of CAD technology is the ability to create automatically a product database consisting of design data, dimensions and key characteristics, as well as associations with other elements of a product. This information is held in an organised structure in the product database and its power

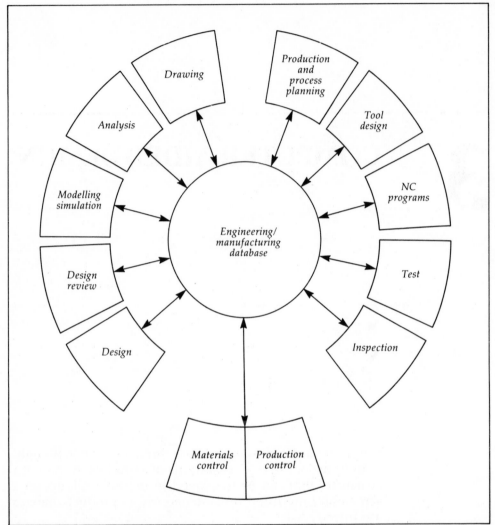

Fig. 8   Role of the product database

is that it can be automatically transformed into a variety of useful outputs – outputs which aid every single aspect of the manufacturing cycle. The role of the product database in CIM cannot be overstressed. This is summarised in Fig. 8, with a specific example from Plessey shown in Fig. 9.

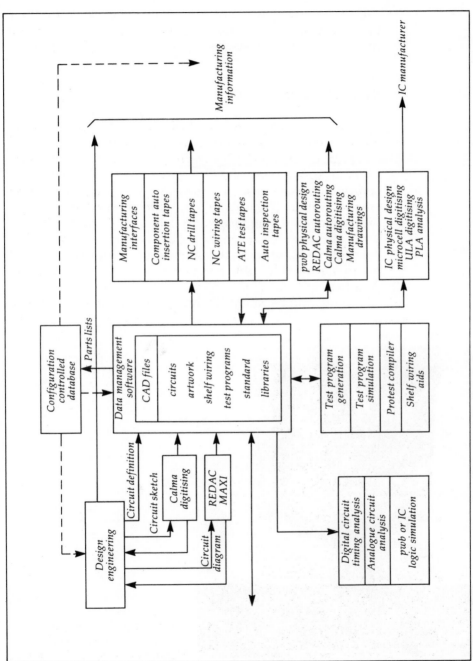

Fig. 9   Computer-aided design system and applications

CAD is one element of the total CIM package that has been very widely applied as a stand-alone function since the benefits it can realise even at this level are substantial.

STAND-ALONE CAD

CAD improves the designer's productivity and quality

By replacing a designer's drawing-board with a CAD workstation, his productivity is immediately increased. In a typical mechanical engineering environment a productivity improvement of 300% can be expected. Since the computer takes care of most of the calculations, the creation of a new design and the modification of an existing design may be accomplished much more quickly. This also results in higher quality designs. With the computer taking over the intricacies of draughting and mathematics (Fig. 10), the designer can concentrate on creative issues. The job is likely to be much more satisfying and he/she is likely to show improved performance.

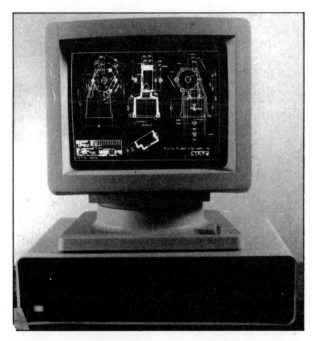

Fig. 10   Complex engineering drawing using CAD (courtesy of Micro Aided Engineering Ltd)

Test and simulation programs create higher quality designs

Higher quality designs also result from the use of test and simulation programs within the CAD system. The notable advantage of higher quality is less design changes and, as a result, reduced lead-time and reduced model-making.

Another advantage of CAD is a reduction in duplication of effort – if all past designs are kept on file they can be readily accessed and easily modified. Thus the designer, generally, need not 'start from scratch'.

INTEGRATED CAD

Once the CAD system is integrated with a full computer integrated manufacturing system, further benefits will accrue. These benefits stem mainly from the fact that the design data can be passed directly via a computer link to other departments in the plant and used as the basis of other activities. Firstly, the design data can be used for process-planning tasks. Secondly, it can be used to prepare the part programs to run the computer-controlled machines which actually manufacture the part. In addition, programs for machining centres, turning centres, sheet-metal cutting machines and coordinate measuring machines can be prepared.

Marketing departments can also benefit from the design data. They can use it to create exploded-view diagrams, schematics, etc., for illustrating technical publications, user manuals, spare parts leaflets, advertising material, and so on. In addition, it is sometimes possible to store a list of standard components and assemblies with their associated costs in the system, so that the design data can be used to produce an estimated manufacturing cost. This is really only successful if a company's products comprise mainly standard elements.

Integrated CAD brings designers and production engineers closer together

The use of CAD as part of an integrated system produces other beneficial 'spin-offs' (resulting in improved part quality) which are difficult to quantify. In the past, the production engineering department has been at the mercy of the designers. It had to manufacture the products that the designers created irrespective of the cost. Traditionally there was a barrier between the two departments and one blamed the other when something went wrong. With the move to CIM, greater communication between the two disciplines is possible. It gives the designers easy access to manufacturing databases so that they design products which meet the required specification and can be manufactured using the facilities available. In this way, production engineering should have very few special cases to handle, and both manufacturing costs and lead-times will be reduced.

Rationalise product designs before taking on CAD

The integrated CAD system should enable a company to reduce its inventory-carrying costs considerably, since it should be possible to use the technology to impose limits on the design department to introduce some design standardisation. For example, if designers are restricted in the thicknesses of sheet metal from which they can select, fewer sizes will need to be kept in stock. If they are limited to a specified choice of hole sizes, drill and tap tools as well as screws can be rationalised. Thus, an extensive rationalisation programme should accompany a CAD installation. This is important not only because of the reduction in inventory but also because of spin-offs such as in FMS where the number of tools, for example, can be a crucial factor due to the large variety of parts one system would handle.

In the electronics field, one of the main benefits of using CAD is the effect it has on the test function. This will be covered in more detail in Chapter Five but the use of CAD linked with computer aided test (CAT) is of the greatest value in electronics manufacture.

HARDWARE

CAD is one of the elements of CIM that have been available the longest; it is also the one for which the greatest number of commercial systems exist. This means that the choice is large and confusing but that well-proven products are available, which is not always the case. The systems available vary from those running on very expensive mainframes, able to support a large number of

Fig. 11   CAD workstation (courtesy of Calma Co.)

workstations, to the microcomputer-based systems with rather limited capabilities.

Whatever the complexity of the CAD system chosen, each designer using it is equipped with a workstation comprising a graphic display and operator input device. This device usually includes a conventional alphanumeric keyboard as well as other equipment such as a function keyboard, a light pen and data tablet (Fig. 11). The whole workstation is designed so that it can be used without any programming knowledge.

Similarly, any CAD system (for CIM purposes) is likely to require a number of output devices to produce designs on paper. A variety of plotters and printers are offered for this purpose. Although one of the aims of CIM is to

Elimination of all
paperwork is the
aim

eliminate all paperwork, it will be a long time before that goal is realised.

SOFTWARE

CAD software is generally structured as a number of separate application packages. The following are examples of some of the packages which are available commercially:

- Mechanical draughting.
- Schematic drawing – standard symbols can be stored on a system and rapidly recalled and positioned on a drawing.
- Design analysis – calculation of properties such as areas, volumes, masses, centres of gravity, moments of inertia, radius of gyration, etc.
- Finite element analysis – meshes for finite element analysis can be produced.
- Standards – standard components and features such as undercuts and tapped holes can be stored on the system and instantly recalled and placed on a drawing in the correct position at the required scale and orientation.
- Parts listing – non-graphical information including part number, material, supplier, weight and cost can be associated with a particular component or assembly. All such information on a drawing or series of drawings can be collected by the system to produce a list of parts.

INSTALLING CAD

It is possible to install a CAD system using workstations from one vendor, a computer from another vendor, printers from a third and software from yet another, but this is most unlikely to give a satisfactory result unless the company has a strong computer programming department.

Use one vendor
and buy a turnkey
system

It really is essential to go to one vendor and buy a turnkey system. This system should be chosen on the basis of the software the vendor can offer rather than the hardware.

The decision about what system to install is an extremely important one because once a system has been established in a company it is extremely difficult and expensive to change vendors. The cost lies partly in the system itself and perhaps more importantly in the value of the data which has been built up, as well as the expertise that has been acquired.

Anticipate future needs for CAD before selecting a system

For this reason, when preparing the specification for the CAD system to be installed the user should anticipate the future CAD needs of his operation to ensure that it can accommodate them.

Software is the only criterion on which to choose a CAD system

Software is the only criterion on which to assess the suitability of the available CAD systems to meet the needs of the specification. Most CAD systems run on a particular range of computers, usually not a very large selection, and it is therefore important to be flexible about the hardware.

Ensure data processing and storage capability are adequate – low response times are frustrating

It may be tempting to cut costs by running a CAD system on an existing computer which is used for payroll or production control, but this is not a good idea. The data processing and storage capacity required by all but the simplest of CAD systems is considerable and the CAD system performance is likely to suffer, giving low response times. Designers will become frustrated with the delay.

The installation of CAD must be considered in the context of CIM and it is vital that the CAD system chosen is able to integrate with the other manufacturing activities. Users must ensure that their suppliers are well informed of the activities relating to the MAP (manufacturing automation protocol) (see Chapter Ten for further details), since this will ensure that an ability to integrate widely will be maintained.

In the move towards full computer integrated manufacturing the need will arise for the exchange of design data between different CAD systems. Design is often an iterative process with designs going to and fro between manufacturers and their customers, manufacturers and their suppliers, or even between different systems in the same manufacturing plant. Traditionally, printed paper drawings have been necessary but greater speed and efficiency is possible when computer links are established.

CAD STANDARDS

Standards for exchange of design information must be developed

At present it is not easy for CAD systems from different vendors to exchange information. However, work on a standard approach to interfacing different CAD systems is in progress and is expected to become accepted internationally.

IGES (Initial Graphics Exchange Specification), was developed by the US Air Force and is now widely used for exchanging 2D and 3D information (including surfaces) and mesh data for finite element modelling. A solids

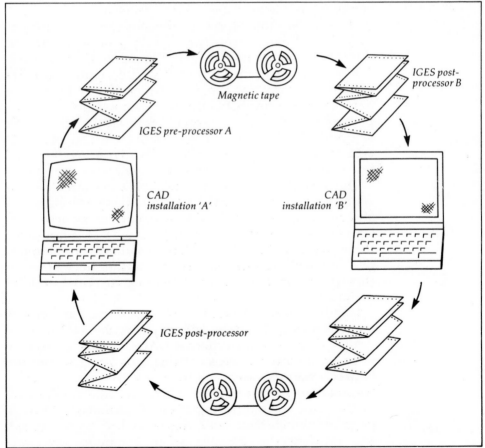

Fig. 12   The principle of IGES

modelling version of IGES is now under development and eventually the specification should cover the transfer of full databases and CAM data.

The principle of IGES is shown in Fig. 12. Each CAD system needing to exchange data with another has to process the data it sends into a standard IGES format and then process the data it receives from the IGES format into a format *it* can understand. Some, but by no means all, of the current CAD system suppliers have made IGES processing facilities available with their products. If this requirement is important to a CIM installation, care must be taken to ensure that the equipment selected has this capability.

Communication between CAD systems and other computerised systems is essential

While IGES addresses the problem of exchanging CAD data between one CAD system and another, it is not concerned with communications between CAD systems and other computerised functions in a plant. This communication is essential, however, when the total CIM environment is to be established.

A number of development projects are underway to standardise the format CAD systems use to define product designs. These are described in Chapter Seven.

4 COMPUTER AIDED MANUFACTURE

Computer aided manufacture (CAM) is the term commonly used for all the computer-controlled activities that are involved in actually turning raw material into finished products. It does, in fact, only refer to production aspects. This chapter therefore deals with the roles of the microprocessor-based controls which are now being fitted to almost every type of processing machine, the means of programming them and their place in the control hierarchy.

Many of the first applications for which computer controls were developed were for metal-cutting machines. Although they are reported to have given a 3:1 productivity improvement over manual machines, the early NC systems had severe limitations. They had no memory, so that for every part in a batch the operator had to feed in the paper tape on which the part program was punched and initiate the machining cycle.

Paper tape itself created problems: tape readers were unreliable and slow, the tape was susceptible to wear and tear, and the programs themselves were often error-ridden. In addition, the early NC machine tool was heavily dependent on the operator for loading the right part and feeding in the right tape. Feedback on machine-tool performance and tool wear was by watching and listening to the cutting operation – a highly judgemental exercise.

NC introduces greater process control

The NC systems available today are extremely sophisticated, providing such a high level of control that multiple machine tools may be run unmanned for extended periods. Whereas controls for metal-cutting machines are still the most highly developed, as are their programming systems, NC systems have been designed for a wide range of processing equipment. Application of NC has been successfully achieved for:

- Machining centres
- Turning centres
- Sheet-metal machines
- Punch presses
- Grinding machines
- Plastics machinery
- Packaging machinery
- Electronic component insertion
- Electronics testing
- Coordinate measuring machines
- Assembly
- Tube bending

The topics discussed in this chapter apply to numerical control in general, irrespective of the equipment it is attached to. However, when reference is made to a specific application, metal cutting is used since it places more demands on the control system than any other application and is currently the most highly developed of all.

DIRECT NUMERICAL CONTROL

Every workstation in a CIM environment must be equipped with numerical control

Every numerical controller in a CIM environment must link up to a supervisory computer

Just as every item of equipment must be fitted with a numerical controller if it is to form part of the CIM environment, so must every NC system link up to a supervisory computer. This communications strategy is often referred to as direct numerical control (DNC).

With DNC, a central computer can coordinate the operation of up to 100 machines. It stores a considerable number of programs, all of which are ready for loading on the machines. Programs are sent to the individual machines as they are needed, which involves real-time two-way communication between the central computer and the machine-tool controllers.

Through the DNC links, data collection can be carried out and factory production monitored. For example, the machine-tool controllers can provide information about the number of parts produced by each machine, machine down-time and tool utilisation. The DNC computer has the task of sorting the data and producing reports for management personnel – thus production problems may be spotted before any bottlenecks occur.

DNC was originally introduced to eliminate paper-tape programs and to enable data collection from the machine tool. The result was a productivity gain of 20% over stand-alone NC machine tools. In the CIM environment, however, DNC is essential. The DNC system should be linked with the part programming systems so that direct transfer of part programs can be effected; it should also connect with the main plant control network in order to participate in the management information function.

In any one plant, several DNC computers are likely to be required, each responsible for a group of machines, either similar or dissimilar, depending on its application. If dissimilar, the machines would be grouped because they process the same part family.

Stand-alone operation of workstations must be possible – computers do break down

It is vitally important that the continued operation of a group of machines is not wholly dependent on the DNC computer, so that production can continue should the computer fail. One possibility is to keep all part programs not only with the DNC computer but also separately, so that in the event of a failure the machines can be run on a stand-alone basis (assuming their NC systems have the facility to accept part programs from an external source). An alternative is to install a standby DNC computer which can take over control should the main DNC computer fail. This solution has been used mainly in the context of flexible manufacturing cells, in which the responsibilities of the central computer are very extensive. The standby computer needs to be in constant communication with the operating computer so that it can assume control immediately a failure occurs. Users of such a solution have been able to use the standby computer for software development, which has helped to justify the expense it incurs.

A standby computer can often by justified

FLEXIBLE MANUFACTURING SYSTEMS

FMS is only relevant in some production environments

The flexible manufacturing system (FMS) can be viewed as a highly complex and sophisticated DNC grouping of machines. While DNC is always part of a CIM implementation, FMS is only relevant in some production environments (Fig. 13).

FMS aims at producing small batches of components (even single components), in a fully automatic mode on a single production line. Every part of the line is computer controlled and can be programmed to produce first one type of component followed by another without manual intervention. A central computer coordinates the operation of all the different elements and links up to the plant's main data-processing computer to provide reports.

FMS is claimed to bring the following benefits: 40% cut in lead time; 30% increase in machine utilisation; 10% reduction in unit costs; and 30% reduction in labour.

If batches are small and part variety is great, machine utilisation is very low, since so much time is spent setting up the machine for a new batch. In FMS, workpieces are set up on standard pallets away from the machine tool and are transported to the machines by an automated handling system, as described in Chapter Nine. Pallet shuttle

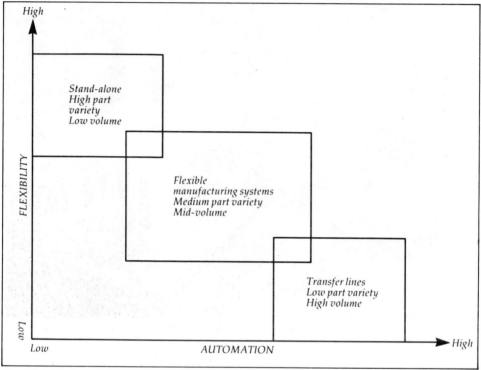

Fig. 13   The application of FMS

devices or robot loading devices transfer the palletised workpieces automatically to the machine's working area and the processing operation starts immediately (see Fig. 14). In this way, the time wasted between the completion of the machining of one part and the starting of the next is minimal. Machine utilisation is therefore very high.

The FMS installation, as described here, is highly automated and is frequently designed to run unmanned for up to eight hours or so. In order to operate in this way it must be equipped with storage for palletised workpieces and cutting tools, both of which are automatically transferred to and from the machine tools by an automated transport system. Automatic identification of workpieces and tools is an essential requirement, as is constant monitoring of the processes.

Fig. 14   View of an FMS plant showing the machining centres, the pallet vehicle and the multi-pallet system

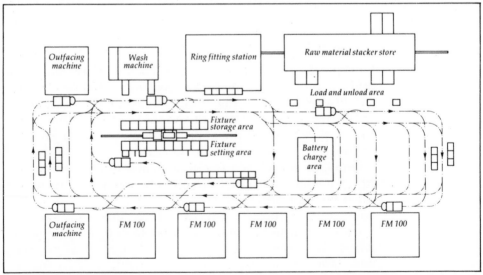

Fig. 15   Typical FMS layout

Computer-controlled machine tools, cleaning machines, coordinate measuring machines and heat-treatment equipment can be combined in a typical FMS installation under a supervisory computer (see Fig. 15). Responsible for coordinating this variety of devices, as well as the storage and handling systems, the FMS computer and its software are of crucial importance if continuity of production is to be maintained even during manned shifts, let alone unmanned shifts.

The FMS computer and its software are crucial

MACHINE CONTROLLER

The CIM environment, whether or not it involves FMS, requires very high performance specifications from the NC systems used to control the processing machines. Fig. 16 shows the role of the machine controller. Some of the most important machine controller responsibilities are listed below:

Machine tool controllers have many duties to fulfil

- *Part identification* – The controller must be able to take information from a sensor, such as a proximity switch or

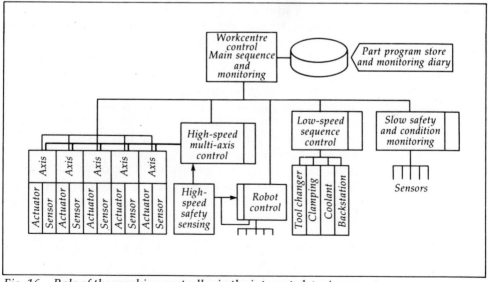

Fig. 16   Role of the machine controller in the integrated environment

barcode reader, to identify the pallet that has arrived at the machine so that it can request the relevant program from the central computer. A special touch trigger probing program may also be used to confirm that the part and program are correct, prior to running the cutting program which establishes the presence of a unique feature.

- *Component alignment* – Probing may also be used when extreme accuracy is required to overcome the cumulative inaccuracies that mechanical fixtures and pallets introduce between the machine and the component. The probe performs surface sensing before machining, and produces fixture offsets which are used as a reference point for the programmed dimensions.

- *Status feedback* – The controller must collect performance data on errors, cycle times, tools, spindle utilisation and parts count, and feed it to the central computer.

- *Tool setting and checking* – The controller must be able to monitor tool life (unique for each tool). In addition, the controller should use data from a tool sensing program involving touching the tool on a table-mounted probe to automatically set the lengths of the tools for machining centres and the offsets for turning tools. This method can also be used for testing tools for breakages.

- *Torque-controlled machining* – In-process monitoring of spindle torque and slide forces should be used to improve cutting performance. The ability to match tool forces against part variations and inconsistencies is an essential technique for improving tool utilisation and achieving a better surface finish. The torque-controlled machining function overrides the programmed feed-rates as the tool becomes dull or as the workpiece material becomes harder.

- *Fault tolerance and tool recovery* – When faults occur and tools are rejected, the controller must be able to effect a recovery and prevent halting production.

PART PROGRAMMING

With the development of increasingly sophisticated numerical controllers designed to aid unmanned manufacture, the part-programming function becomes more and more difficult. It is a function which is closely interlinked with the design and, whether programming a five-axis machining centre or an electronic component insertion machine, the starting point is the product design held in the CAD database.

Computer aided programming is invaluable

It is not yet possible to automate the part-programming task. No one has yet set up a computer-aided part-programming system which can turn a design into a program without a sizeable amount of programmer intervention. However, there are many computer-assisted programming systems available which are essential in any CIM implementation. They speed up the programming process quite considerably, reduce the possibility of programming errors and, in addition, can be used by people without an in-depth knowledge of computing and software.

Some CAD/CAM systems are intended for both designing and programming to be carried out using the same computer. In this way, a user who has previously used an outside specialist could venture into in-house part programming. However, any user should separate the two functions onto two separate computers at the same time, ensuring that they integrate easily.

Sub-routines reduce the programmer's burden

In the past, the most widely used machine-tool programming language was APT (automatically programmed tools). This is suitable for positioning and continuous-path programming in five axes. However, in the drive towards integration and automation, considerable development of programming languages is taking place, aiming at simplifying the task for the complex manufacturing systems being planned. The development is particularly active in the robotics sector, and many different solutions have been proposed. It is particularly important that, whatever language is used, sub-routines are included. Using sub-routines for those motion sequences which are repeated several times during a program, the number of statements required in the program is reduced, making the programmer's job less time-consuming.

In preparing a part program, the programmer first has to access the part design. With the interactive graphics provided with most available programming packages, the programmer can view the design geometry on a screen; he can also manipulate the design and view it from different sides, if necessary. The next job is to label the various surfaces and elements of the design. The system will then automatically generate the APT geometry statements.

Simulate the work process before finalising the program

To continue with the programming function, the programmer needs to access a CAD/CAM database containing a listing of tools, machines, recommended feeds and speeds, etc. He must select a tool from the database and then proceed to plot the tool path. Using the programming package, the tool path can be generated in an interactive mode. Facilities should be available to simulate the tool path on the graphics screen and overlay it on the design, so that the programmer can verify it and check that no collisions between tool or tool-holder and fixture occur. Because of the ease of generating a tool path, the programmer can try out different approaches, using different tools, and feeds and speeds, until he/she is convinced that the optimum result is achieved.

POST-PROCESSING

The tool path generated on the computer-assisted part-programming system is sometimes known as the CL (cutter location) file. It is not yet specific to any particular machine tool. To convert the CL file into a part program for a particular machine tool it must first be post-processed. The post-processor is a program containing all the information relating to a particular machine tool and control system combination which converts the CL file into machine tool code. (Packages for developing post-processor programs, in a conversational mode, are available but this is an area where there is still some room for development.)

Post-processing should be carried out at the last possible moment for optimum flexibility

Post-processing can be performed at various different stages in the programming activity. Most flexibility is gained if the post-processing is carried out at the latest possible moment when the status of the various possible machines is known most accurately. Nothing is achieved by dedicating a program to a machine which is undergoing maintenance, for example. In fact, post-processing is sometimes included in the computer-assisted programming software. One other possibility is that the machine controllers offer the facility to post-process CL files as they are being down-loaded.

MANUAL DATA INPUT

Machines are currently being offered with sophisticated controllers which enable direct programming on the shop-floor. This facility is known as manual data input (MDI). The products offer many attractive features such as conversational programming with elaborate graphics but they are not really of relevance in the CIM environment.

5 COMPUTER AIDED TEST

With the growing complexity of electronic circuitry new technology for testing has been introduced. Using computer control, this new technology, known as computer aided test (CAT) has been made possible by further developments of the integrated circuit (IC).

The integrated circuit presents a challenge to test engineers because each device performs a very large number of functions, and their performance is still increasing. Whereas in 1960 a single computing function was placed on a semiconductor chip, today there are ICs with more than 400,000 functions. And there are forecasts of up to 10 million functions per IC by 1990.

Effective IC testing simplifies testing of pcb assemblies

To test a single integrated circuit, each function must be analysed, and all possible faults must be identified and located. This is no easy task, but when a number of ICs are assembled with other components on a printed circuit board (pcb), the test function is immensely complex. If the IC test is carried out effectively, then the subsequent test stages – those of an assembly of pcbs and a complete system – will be that much easier.

Find faults at the earliest possible stage

It is vital to find faults at the earliest possible stage in the manufacturing cycle, since the cost of repair increases by a factor of ten at each test stage. Therefore, the cost of repair of a pcb is ten times the cost of repair or replacement of one of its components. Many companies prefer to put the onus for component quality on their suppliers and pay the higher cost to guarantee a certain standard. The alternative is to test components in-house but the equipment to do this is costly and difficult to justify unless very large volumes are involved, and can lead to large number of ICs being discarded.

This chapter concentrates mainly on the stage of electronics testing concerned with pcbs, partly because it is the most crucial, and partly because it is the one which integrates most closely with the design process.

Printed circuit boards are usually designed to perform a specific purpose. They contain a number of standard components, including both static components, such as resistors and capacitors, and dynamic components, such as integrated circuits. Each device has performance characteristics associated with it, as does each assembled pcb. It is therefore possible to test whether a certain pcb is good by giving it certain input signals and analysing the output. With the 'very large scale integration' (VLSI) circuits currently being manufactured, a single pcb can be extremely powerful, and the input signals to test its performance characteristics and investigate every possible defect that could occur are highly complex and time-consuming to develop. Further complications are introduced if there is the requirement to locate the fault precisely. The set of signals required to test a complex pcb, known as the test program, is relatively easy to develop and swift to run, but a fault-finding program is not only

Fig. 17   Automated test station (courtesy of Everett/Charles Test Equipment Inc.)

difficult to generate but also lengthy to carry out. The dynamic nature of many VLSIs means that very high speed test systems are required and their generation and detection makes fault isolation more difficult.

In fact, so lengthy and complex is the testing of pcbs that without computers it would be impossible to do effectively. The use of a computer to generate, implement and analyse test programs, speeds up the entire test process, thus enabling more exhaustive tests to be carried out. It also gives the tests a very high degree of accuracy (Fig. 17).

INTEGRATING DESIGN AND TEST

Design and test disciplines must work in close collaboration

As mentioned earlier in this chapter, the design and test of pcbs should be closely integrated. The computer plays a very important part in bringing the two functions together. In the past, the testing of a component or board was not considered until the entire design process was complete. Design and test was the responsibility of two different engineering groups, between whom interaction was mostly non-existent. There was therefore a good chance that the design created was difficult or even impossible for the test engineer to test.

The move towards integration does not eliminate the need for both design and test engineers with their respective disciplines, but it must involve the design engineer in testability considerations. It must also set up the essential and previously ignored lines of communication. The result is pcbs which take less time and therefore less cost to test, and thus improve the reliability and marketability of the product.

Simulation must be used in design to ensure pcb can be tested

The designers of pcbs must have CAD tools at their disposal. Part of the CAD software includes a simulator package which often can automatically generate a test program for the circuit being designed and enable the designer to analyse its testability. The designer should therefore be responsible for modifying the design in an iterative mode until the simulator indicates that an effective test program can be generated. The simulator software can model the number of possible defects that occur in the circuit and compare this number with the number of defects the test program is able to find. A designer should not be content with the design until the test program is capable of finding at least 98% of all faults.

TEST PROGRAM GENERATION

The test program forms the basis of the entire procedure

The test program generated by the simulator software then forms the basis of the entire test procedure. Firstly, it can be used to run the computer-controlled test equipment which applies the appropriate test signals to the relevant connections on the pcb, and analyses the output against the expected characteristics.

The test program, as generated by the simulator, often cannot be fed straight into the test equipment. It must first be post-processed into a form acceptable to a specific test machine. This will always be necessary because of the large variety of vendors of design and test hardware and software. Post-processing software for certain combinations of equipment is readily available; if it is not available, it has to be specially written, but this is expensive. It is therefore preferable to minimise the number of different CAD systems and CAT equipment.

Standardisation on
test equipment is
not yet possible
It is not really feasible to try to standardise on test equipment, since this has also been going through a similar revolution as integrated circuit design. Every year new equipment becomes available and the different types of machine all have some desirable attributes. Speed is the important feature of some; others have useful electrical characteristics. In some applications, simplicity of the device adaptor is relevant, and the user-friendliness of software varies from machine to machine.

One type of machine, favoured in the past, uses normal plugs and sockets around the edge of the board to connect it to the test system. However, with the introduction of VLSI devices and the intense test programs they require, this method cannot give adequate access to different parts of the board, especially for fault finding.

Fig. 18   Bed of nails fixture (courtesy of Everett/Charles Test Equipment Inc.)

To provide the access needed for complex pcbs, the 'bed of nails' technique has been developed. Now well established, this method uses a fixture consisting of a bed of spring-loaded probes to connect the pcb to the test machine (Fig. 18). Each probe is connected to its own piece of electronic circuitry (node) which allows that pin to be used for either stimulus or measurement. In this way, every single connection and component on a pcb can be accessed in the test program, if required.

The matrix of probes on the 'bed of nails' fixture will be different for each pcb to be tested. Again, data to create this matrix can be extracted from the pcb design information produced by the CAD system. This data may be required as a paper print-out showing the pin layout, so that it can be used as a mask to construct the fixture. Alternatively, it is possible to output the data in the form of an NC tape and use this to control a drilling machine to drill the pin holes in the required places in the 'bed of nails' fixture.

The 'bed of nails' technique has in the past had some reliability problems, mainly mechanical, but also caused by cross-talk between closely spaced pins. These problems have generally been solved; the only real limitation now is that on each fixture there is a maximum number of pins and this may not always be enough for a very complex board.

BENEFITS OF CAT

CAT should be integrated with design and production control

It is clear that there are a great many advantages to be gained from using computer aided test in an integrated manner. The data created in the design process can be used not only to control the test equipment but also to design and construct the fixture to hold the pcb in the test machine. An additional benefit is that the test machine can collect data about the boards going through the system and feed it into the integrated system, so that the production control department can update its records. With a link between the main host computer and the test stations, statistics on failure rate and types of failure can be maintained. With this information used correctly, it is possible to make continual improvements to the process. For instance, if components are faulty or supplied with bent leads, the suppliers can be notified. If components are incorrectly orientated on the board, not inserted properly or repeatedly totally absent, there are obvious improvements to be made to the assembly process.

CAT must not encourage assembly machine operators to take less care

There is no doubt that CAT plays a vital role in electronics manufacture but its implementation can be problematic and should be handled with care if the full benefits are to be realised. One risk associated with the installation of CAT is that the operators on assembly stations may feel that the quality of their work is not so crucial if 'magical' machines are used to pick out all the defects. If this happens, they are likely to become less careful and there will be a bottleneck at the test equipment while the machines spend time locating faults. In addition, the level of rework will rise. Total integration would also automate the assembly and repair stations.

SELF-TESTING

Be aware of built-in test as a future feature of pcbs

The future could see quite dramatic changes taking place in the domain of electronics testing as built-in test becomes a feasible possibility. The incorporation of a self-test capability into pcbs, composed almost entirely of large scale ICs, is a viable and powerful alternative to testing solely controlled by external test equipment. In self-testing, a microprocessor in a circuit board sends out a pre-programmed set of signals to other components on the board, instructing them to send certain signals back. By comparing the returned signals with those expected, the microprocessor can decide if the circuit board is functioning properly.

The board with in-built test will be designed with a special controller. This will probably manifest itself as an extra chip but it could be incorporated into one of the devices on the board. The level of complexity of the controller depends on the required level of test. If only a 'go/no-go' indication is required then a simple-state

Disadvantages
- *Input-output data comparison is complex*
- *Self-test control circuitry is not generally tested*
- *Needs additional circuitry which is redundant from a systems-function point of view. Real estate on chips is expensive*
- *Failure of self-test facility invalidates test*
- *Additional software required*

Advantages
- *Saves time*
- *Reduces operation/support expenses*
- *Speeds-up field service testing*
- *Minimises test access and support requirement*
- *On-board self-test allows microprocessors to be tested at normal operating speed*
- *Provides an integral diagnostic capability*
- *Does not require removing LSI/VLSI board*
- *Does not require separate CAT*

Fig. 19   Advantages and disadvantages of self-test

machine may be used as the test controller. If, on the other hand, a diagnostic level of test is to be achieved, then the controller may be a simple processor.

The advantages and disadvantages of self-test are shown in Fig. 19. Self-test may add expense to the construction of a pcb but the benefits can far outweigh the additional costs and it is likely that in future all electronic systems will become self-testing.

SURFACE MOUNT DEVICES

The use of SMDs poses more demands for the test function

In the more immediate future, however, the use of surface mount devices (SMDs) is to become more widespread. These will place new demands on test technology because the closer proximity of components presents increased difficulties in locating test pins for 'bed of nails' testing. However, if design improvements are achieved as anticipated, then some stages of the test cycle may be bypassed.

Studies of the feasibility of testing SMD circuit boards particularly in high reliability environments have begun to yield results. British Telecom Research Laboratories (BTRL) researchers, for example in manufacturing a trial echo-canceller circuit in prototype form to evaluate surface mount technology, have found potential problems in the extremely close spacing of the components. They conclude that to avoid them it places a heavy responsibility on the circuit designer to design-in test options at the beginning, in exactly the same way as custom integrated circuits. Metallurgically, surface mount technology may prove to be less reliable than standard moulded dual-in-line packs, because at present they are expected to survive a total immersion in hot solder as part of the manufacturing process – which does not apply to conventional packaging, although high reliability boards have been demonstrated.

6 MANUFACTURING PLANNING AND CONTROL

The planning and control of manufacture relates to the organisation of the production facility so that customer orders are handled efficiently and economically and delivered on time. This involves organising the purchase of raw material and the start of the various production processes so as to meet delivery dates. It also involves making allowances for all those unpredictable elements which can cause chaos even to the best planned activities.

Customer orders must be handled efficiently and economically and delivered on time

Available machinery and manpower resources must be monitored to ensure that production requirements can be met, and product costs must be kept under strict observation. The tasks of this function are therefore many and varied.

During the last ten years or so, attempts have been made to formalise the tasks involved in manufacturing planning and control. They were previously considered to be no more than clerical jobs and there was little coordination between them. In addtion, there were few rules or methodologies as to how best to perform the different activities. Planning and control was therefore extremely haphazard.

Profitability depends on effective planning and organising

However efficient the manufacturing processes in a company are, the company will not be successful and profitable unless it organises and plans its activities effectively. For example, production schedules will not be adhered to if they fail to take into account the availability of equipment and personnel resources. This will result in failure to meet delivery deadlines, customer complaints, and back-ordering or excessive overtime.

In the past, planners have been tempted to cover for any eventuality or emergency to ensure that delivery times are met. They assign unnecessarily long lead-times to components and allow massive inventories to build up – both useful ploys in case problems occur and both very costly and bad practice for any manufacturing concern interested in profit.

The manufacturing planning and control function has developed considerably in recent years, and now has systems, methodologies and a terminology of its own which have promoted it to the professional status such an important function deserves.

A planning and control system is individual to each company

Different companies have defined the varied tasks associated with a total manufacturing planning and control system in different ways. The following is one company's proposed elements:

- Inventory control
- Bill of materials
- Master production scheduling
- Materials requirements planning
- Capacity requirements planning
- Shopfloor control
- Job costing
- Customer order entry

- Forecasting
- Financial resource planning
- Accounting

Each of these elements has been computerised and each on its own can be beneficial, but the most significant advantages are gained when the elements are integrated. Much of the data generated by one element is needed by another, providing the perfect opportunity for the introduction of computer integration. Fig. 20 shows the

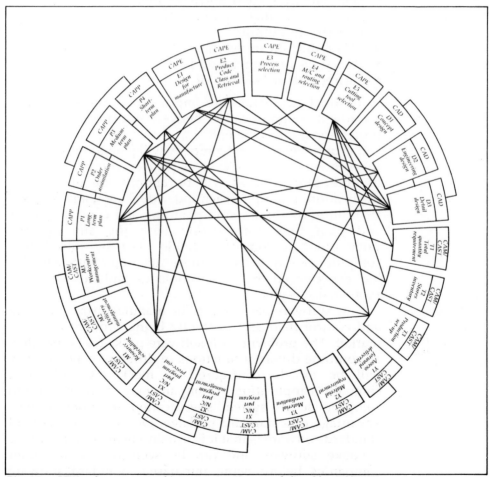

Fig. 20   CIM subsystems and interconnections as defined by ESPRIT

subsystems and interconnections as defined by the ESPRIT project on design rules for CIM.

Rules and methodologies must be precisely defined and adhered to

The recent application of computers to the production planning and control function makes it easier for the rules and methodologies to be defined precisely for a particular company and then to be adhered to. The importance of computerising the manufacturing planning and control function is shown by the following demonstrated benefits:

- 20–33% reduction in inventory
- Manufacturing costs down 10–15%
- Inventory shortages reduced by up to 80%
- 10% rise in output
- 50% reduction in overtime
- 5% savings in purchased parts
- Indirect labour savings of 10–25%

Additional advantages are improved customer relations, improved communication between functions and more professional management.

Tackle the worst areas first

It is unlikely that any company will plan to introduce all the elements at once or even all the elements over a period of time, since not all will be appropriate in any one application. It is up to each individual company to select its unique combination of planning and control systems and plan their introduction with the aim of total integration. A company should start by introducing computerised systems for those functions with which it has experienced the most problems, and this will give immediate results.

Standard software may be a compromise but it's cheaper and more reliable

New software packages in the manufacturing planning and control field are constantly being brought onto the market. The problem with software packages is that they all combine different functions in different ways but few will be exactly configured to meet any individual company's application. A company may, therefore have to compromise its system specification in order to take advantage of a standard package. One of the benefits of standard software is that it is usually cheaper than writing bespoke software and can be seen in use by other companies before being purchased. Standard software will also take less time to introduce.

Adopt 'just-in-time' (JIT) as a target

Before a company makes any decisions about which elements of the total manufacturing planning and control function it is to computerise first, or which software it will purchase, a rethink of general production organisation should be carried out. The move towards CIM must be accompanied by a change in philosophy and method of organisation, to approach the 'just-in-time' (JIT) way of working.

JUST-IN-TIME

JIT is a 'pull' approach driven by final stage of assembly

Originally known as 'Kanban' when it was developed by Toyota in Japan, the principal of the just-in-time philosophy is that every process in the production cycle happens just in time for the next. For instance, raw material arrives at the receiving department just in time to go to the first machine for processing; final assembly and test of products is carried out just in time for dispatch to customers; and so on. This is an ideal state of operating and, although it will probably never be achieved, it is a very good target to have in mind when implementing elements of manufacturing production and control systems.

To operate JIT principles, raw material and components should only be purchased as needed to satisfy customer orders – the economic order quantity is redundant. The main benefit is in the reduction of inventory: finished goods, work-in-progress and raw material. In addition, quality improvement is a side-effect and with less work-in-progress, if a problem is discovered there is less rework to be done.

MASTER PRODUCTION SCHEDULE

The 'master production schedule' (MPS) is the basis of the entire manufacturing process because it comprises a definitive statement of which products, in what quantities, are to be produced over a period of time, allotting a due date to each batch.

Some companies may produce to stock only, others to order only, but most work to a combination of customer orders and projected sales. The MPS has therefore to take into account both types of work and operate in close collaboration with the business planning function and the sales department.

The master production schedule must be realistic

The importance of the MPS is paramount as it forms the basis of many decisions involving the purchase of raw material and the commitment of production equipment and labour. If it is not realistic (and the tendency is for companies to be over-optimistic about lead-times and the number of orders that can be fitted into any time period), the consequences are costly. The result is that excessive inventory (both raw material and work-in-progres) builds up. It is therefore essential that the MPS is based on good information and that it operates closely with the capacity planning function which analyses the effect of the loads projected by the MPS on specified key resources such as machines, cash flow, labour and space.

The MPS is almost certain to change during its life-cycle. New orders will be received, existing orders cancelled, and the priority of orders changed.

BILL OF MATERIALS

Each item on the MPS has to have a 'bill of materials' (BOM) file created for it. The BOM lists all the raw materials and components that are required to make a product. The process of creating a BOM is sometimes referred to as an 'explosion of parts'. The product is 'exploded' into its constituent parts, as shown in Fig. 21. The subassemblies that make up a product, the components that make up these subassemblies and then the raw materials required to produce these components, are all

The CAD database should be harnessed to produce the bill of materials

identified in turn. This process is not difficult to perform and is most efficiently handled through direct access with the CAD database where the product and component designs are held.

The BOM is used in several of the subsequent manufacturing planning and control functions. It forms

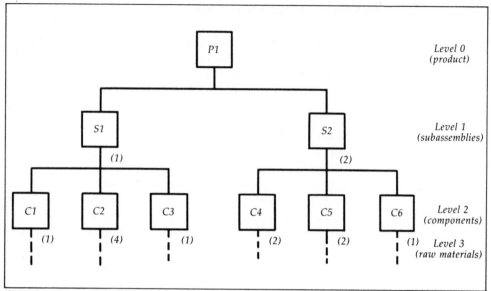

Fig. 21   'Explosion' of product to create a bill of materials (subassembly S1 is the parent of components C1, C2 and C3; product P1 is the parent of subassemblies S1 and S2)

the basis by which raw materials and components are ordered so that an item on the MPS can proceed. It also forms the basis for withdrawing material from stores for manufacturing. In addition, used in conjunction with the routing file or process plans, the BOM makes it very easy to cost products.

By computerising the BOM file, it ensures that the design, production and accounts departments all work to the same list of parts. It also provides the ability to search the total product list to find out which parts contain a specific component or which components a specific product contains. This can be important if trouble arises with a particular workstation, bought-in component or order priority. The effect the problem has on production can therefore be looked at very closely.

The 'bill of materials' (BOM) must be able to cope with change

The BOM is usually very static but systems must be able to cope with the changes which will periodically be necessary.

INVENTORY MANAGEMENT

The management of a company's inventory is an important task since mismanagement can lead either to excessive stocks of useless items, or to stock shortages which hold up production or prevent sales. The aim is to minimise investment in inventory, while at the same time maintaining a sufficient level to provide customers with a good service.

Inventory management includes inventory accounting and inventory planning

Strict inventory control is needed for raw materials and purchased components, work-in-progress, finished product and production peripherals such as tooling, and spare parts for machinery. The inventory management module should be responsible for two jobs: inventory accounting and inventory planning and control for all these forms of inventory.

Inventory accounting involves keeping track of what inventory is resident in a company. Contained in an item master file or inventory record file, the data against each item should include details such as lead-time, cost and order quantity as well as a time-phased record of inventory status, and miscellaneous information about purchase orders, scrap and rejects. The source of this kind of data is a network of data input terminals spanning many different departments such as goods receiving, raw materials stores, tool stores, dispatch, etc.

There are two methods of maintaining inventory levels. The traditional procedure is to set a re-order point for each item on the master file. This involves determining at what level of stock new inventory should be ordered and what quantity of goods should be purchased in each re-order. This method is still suitable for finished product and production peripheral inventory. For the other forms of inventory – raw material and purchased components and work-in-progress – material requirements planning (MRP) is becoming widely used.

MATERIAL REQUIREMENTS PLANNING

While the re-order method of inventory control is based on past usage of items on the master file, material require-

*MRP is a 'push'
approach driven
top-down by
external demands*

ments planning attempts to organise inventory replacement according to the future demand for products. In this
way, it should generate orders more closely in line with
actual requirements and hence reduce the possibility of
excessive stocks or shortages.

In order to plan the material requirements, MRP needs
to access the master production schedule, the bill of
materials and the inventory record. From the MPS it
finds out which products have to be produced and
when. The BOM then enables the schedule to be
converted into a requirement for materials – both in-house
manufactured items and bought-in goods. By comparing
this requirement with the current inventory status, MRP
can calculate what new stock of raw materials and
bought-in components needs to be purchased in order
that the master schedule can be achieved. The structure
of a materials requirements planning system is shown in
Fig. 22.

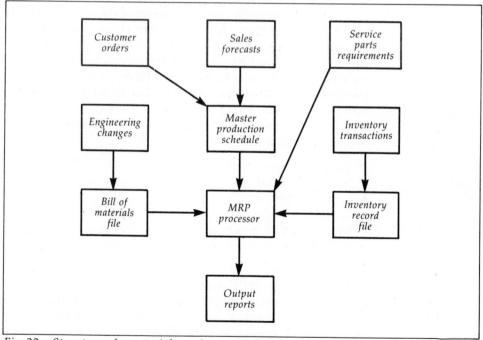

Fig. 22   Structure of a material requirements planning system

The MRP system should not just be able to compute material needs but should also work out the timing for new orders. By working backwards from the delivery date, and by consulting process plans for manufacturing times and inventory files for lead-times on bought-in items, the MRP system should be able to define when orders for raw material and bought-in components need to be placed. Thus, the output from the MRP system is a series of re-order release notices. In generating these notices, it is sensible to collect together any raw material and components common to many products and combine them into a single net requirement.

A single run of an MRP system can take some time (sometimes as long as a weekend) if the files are big and the products complex. As a result, a company may only run its MRP system at infrequent intervals (no more than once a week), whereas data relating to inventory status is constantly changing. To be able to react more quickly to the changing situation, a form of MRP has been developed which involves much less processing of data and therefore takes much less time. More frequent running of the MRP system is therefore possible. Known as 'net change MRP', this faster method of materials requirements planning is only concerned with changes that have occurred since the last time the MRP system was run. The more frequently the system is run, the fewer the changes there will be to handle and the less time it will take to run. The changes which occur are in the following areas:

- Product requirement
- Bill of material
- Item master file
- Orders raised
- Movement of goods in or out of company
- Movement of goods in or out of stores
- Work scrapped

The net change MRP system looks at the file of changes that have occurred since it was last run and effects the necessary replanning of requirements and orders. This system is of course highly dependent on the constant feedback of data from the shopfloor so that accurate information is being used at all times.

Good vendor relationships are critical to efficient planning and control

It has always been assumed that vendors can deliver parts and raw material of the required quality at the required time and in the correct quantity. This is not in fact the case because vendors suffer from exactly the same problems as the manufacturers. It is essential for there to be a good customer/vendor relationship, almost to the point of co-existence. To establish this good relationship it is fundamental to share the master schedule and agree changes in lead-times, to provide rapid feedback on problems, share strategy on future requirements and to show that the vendor is valued by paying promptly and keeping to other contractual conditions.

7 PROCESS TECHNOLOGIES

Process planning, the activity linking CAD and CAM, is possibly the least advanced element in the CIM environment. The task of defining the process by which a product is to be made is, however, a crucial part of the manufacturing cycle. For every product there are numerous methods, some of which will be better than others. So far, computers have been employed more to improve the consistency of process plans than to automate their formulation. As this is traditionally a very labour-intensive area, the potential for improvement is great.

Fully automated process planning is a goal for the future – it is not yet possible

This chapter looks at how computer aided process planning is developing and how coding and classification systems such as 'group technology' can play an important part. Also outlined is the research on product data definition and expert systems which will one day enable fully automated process planning to become a reality.

Every new product needs a process plan

A process plan is required for every new product to be manufactured. This is a plan outlining the specific manufacturing details and is an extensive document which is likely to contain some or all of the following information:

- The raw material to be used to manufacture the part with appropriate data about its form, dimensions and condition
- The operations to be performed, the machines they are to be carried out on and the sequence in which they are to be done
- Tools, jigs and fixtures to be used in association with each operation
- Cutting feeds and speeds for each of the machines
- Process parameters such as temperature, vacuum, etc.
- Inspection requirements
- Standard times for the operations to be performed as well as the time for changeover

Once this information has been established, there are many uses for it in many different departments. For example, the programming department needs the process plan to establish which machines to program which parts for. The scheduling department needs it to allocate jobs to machines, and the estimators need it to calculate product cost.

The process plan must be made available to many different departments

In the ideal CIM environment, the design data created at the CAD facility could be automatically converted into a process plan which would then be held in a database for all interested departments to access as required, as shown in Fig. 23. Such an eventuality is not an impossibility but it is some way off in the future. Before it can happen, two major development areas will have to be solved: firstly, the way in which product design data is defined needs to be clarified; and secondly, expert systems have to be developed to replicate the decision making process that the manual process planners have used in the past. Both of these topics are covered later in the chapter.

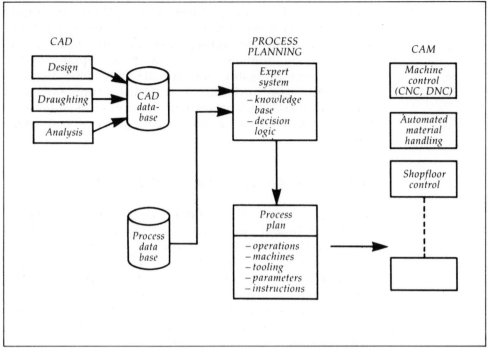

Fig. 23   Integration of CAD and CAM through knowledge-based process planning

MANUAL PROCESS PLANNING

The process planning activity involves several highly-skilled and decision making aspects, as well as tedious data retrieval and data processing jobs. To start with, the planner looks at the design and, on the basis of his/her experience in manufacturing, assesses how it should be made. This is the skilled part where detailed knowledge of materials, machines and products is essential. Tables of data listing those machines and their specifications available in the company, raw materials kept in stock, tools, etc., assist the planner in completing the plan. Accessing this kind of data and performing the necessary calculations is the time-consuming and tedious part. Finally, the process plan has to be set out neatly in tabular form.

The manual process planning task is, therefore, highly labour-intensive and requires high skill levels, although much of the time-consuming part of the job is routine. It is also extremely judgemental – no two process planners will ever produce the same plan for the same part. Benefits would, however, undoubtedly accrue if a company could ensure all process plans were of the optimum standard – one advantage of using computer aids in process planning.

COMPUTER AIDED PROCESS PLANNING

Computer aids available today reduce the labour intensive aspects of process planning but still require skilled operators

Other advantages of computer aids in process planning are the increased productivity of the planners and the improved presentation of the process plan documentation. The computer aids which have emerged so far have mainly succeeded in reducing the tedium of the job. The commercially available packages virtually all consist of a good, comprehensive database and some word-processing ability. They fall into two categories: variant and generative.

Variant process planning

The variant form of computer aided process planning (CAPP) relies on the fact that the parts produced in any one company bear some resemblance to one another and to parts produced in the past and those expected to be produced in the future. Most of the parts a company produces can usually be split up into part families in which they are grouped according to their manufacturing characteristics. For each part family a standard process plan is established and then stored in a database. Having set the system up in this way, all the planners need to do when a new part comes along is to establish which part family the new part belongs to and then to modify the standard process plan for that family to tailor it to the new part.

For this form of CAPP to work effectively, careful coding and classification of parts is essential so that new workpieces can be quickly allotted to a family and so that the required standard process plans can be easily retrieved. (Coding and classification is discussed later.)

CAPP increases the consistency of process plans

Variant CAPP is still heavily labour-intensive but it should reduce process-planning times quite considerably and ensure that a more consistent high quality of plans is achieved.

Generative process planning

The generative approach to CAPP sets out to develop a process plan from scratch from the component drawing. Most commercially available packages in this category are interactive, relying on an operator to input detailed information about the component geometry every time a new component is required. Although this is time-consuming, the packages do enable users to build their own logic into the system and this should eliminate much of the expertise that is a prerequisite of manual process planning.

Generative CAPP may also employ some form of coding a classification of parts to summarise part design information, as this would reduce the time taken to input data. In practice, the commercially available packages of this kind are only suitable for a fairly limited number of manufacturing processes. Since manufacturing logic is highly variable from one situation to another, existing generative systems are very specialised systems developed to suit a specific set of operations performed in a specific manufacturing environment.

GROUP TECHNOLOGY

Group technology is an aid to reducing unnecessary duplication of work in many fields

In the late 1950s, the Russians first recognised the benefits of making use of the similarities between parts. In recent years, this concept, known as 'group technology' (GT), has become widely used in different applications. Areas such as design, process planning, inventory ordering and control, and part manufacture can benefit from the implementation of a GT system. Consider for example, a company which handles thousands of part numbers. Without a coding system this company would probably find it quicker to design a new part from scratch than to look through its existing list of parts for an identical or even similar part; this task, however, is simple with a coding system.

Parts must be coded and grouped

According to the type of GT application, parts can be grouped in different ways. Members of a group may share similarities either in their design or in the way in which they are manufactured. The task of identifying the groups and their members is a long and involved operation. The company must look at the individual design and/or manufacturing attributes of each part, each attribute being uniquely identified by a code number. Typical attributes that may be included in a company's group technology classification system are shown in Fig. 24.

Two basic types of coding have been used in the GT context: hierarchical and chain-type. A hierarchical code is one in which each digit in the code depends on the meaning of the digit which precedes it, whereas the digits in the chain-type code have a fixed meaning. In fact most of the commercially available coding systems are a combination of the two. Hierarchical codes tend to be able to communicate more information with fewer digits than chain-type codes which can become rather lengthy.

Part design attributes	Part manufacturing attributes
Basic external shape	Major process
Basic internal shape	Minor operations
Length/diameter ratio	Major dimensions
Material type	Length/diameter ratio
Part function	Surface finish
Major dimensions	Machine tool
Minor dimensions	Operation sequence
Tolerances	Production time
Surface finish	Batch size
	Annual production
	Fixtures needed
	Cutting tools

Fig. 24   Design and manufacturing attributes typically included in a GT classification system

Computer aids, operating on an interactive basis to develop the codes, are commercially available.

Once a company has coded its complete list of part numbers, the opportunity exists to search parts which are similar in many different ways. A number of applications of group technology are outlined in the following section.

GT and part manufacture

Group technology first found popularity with the GT cell – a set of dissimilar machines which are grouped together so that they can complete the processing required for a group of parts. This differs from a traditional plant layout where all turning machines occupy one area, all milling machines another, and so on. In a GT cell, a variety of parts, which may be of very different design but require similar manufacturing processes and fall within the same size envelope, are produced. This variety, or family of parts, can be identified using the type of coding and classification system mentioned above. In creating a GT cell it should be possible virtually to eliminate the set-up times for new batches, thus introducing to batch production the economies of scale possible with mass production.

Re-organising for group technology does require investment

To set up a system of GT cells requires an investment in part coding, staff training and in rearrangement of machines, but the benefits that can result are extremely significant:

- Reduced set-up time (up to 60%)
- Reduced tooling costs (10–40%)
- Reduced work-in-progress (up to 50%)
- Reduced throughput time (up to 60%)
- On-time delivery
- Increased worker satisfaction

GT and design

It has been estimated that in a typical company 40% of new parts are identical to existing ones, 40% can be designed by modifying existing ones, and only 20% require design from scratch. Group technology can therefore assist the design function quite substantially. A company which has hundreds of live part numbers on its books and which handles several new ones every week should not hesitate to set up a coding and classification system. If every existing part is coded according to its design attributes, the designer should be able to code any new part quickly, access the GT database for identical or similar codes and retrieve the designs. The existing design can then be modified if necessary, which would take considerably less time than starting from scratch.

In addition to reducing design time, the use of a GT code can help to improve on design standardisation – a benefit which has numerous spin-offs.

GT and process planning

Development of data definition standards and expert systems will eventually make it possible to construct a fully automated process planning facility – but not for some years

As explained earlier in this chapter, group technology is the basis of most computer aided process planning systems currently available. GT, however, has not yet provided the means to attain the ideal situation – the fully automated process planning system, working straight from design data developed on a CAD system. The many research and development programmes currently tackling this task focus on two areas: design data definition and expert systems.

DATA DEFINITION

Computer aided design has found very widespread application, but it is difficult to use the part representations obtained with most current CAD systems because they give insufficient attention to non-geometric data. For process planning a feature-based description is required plus details of dimensional and surface tolerances.

New modelling systems are currently being developed which are feature-based and these should create data in a suitable format – but not necessarily a standard format. To speed up the development of total product data definition, the National Bureau of Standards and the US Air Force are supporting the development of a Product Data Exchange Standard. This aims to develop one model of the product in digital form which can be used by all the functions over the life-cycle of the product. In Europe a programme known as 'Standard for the Transfer and Exchange of Product Model Data' is also engaged in this work. The outcome should advance the progress of CIM development considerably.

EXPERT SYSTEMS

An expert system is a rule-driven system which seeks to emulate the reasoning capacity of an expert in a particular area. It has considerable advantages over conventional structured computer programs in applications such as process planning for a number of reasons. For example, an expert system offers a modular architecture for building large programs, and knowledge in the form of production rules may be added, deleted or modified in the knowledge base without any alteration in the control structure. In addition, expert systems can acquire knowledge interactively through dialogue with a user and have the ability to explain the line of reasoning used in any particular situation. They can also perform symbol manipulation so CAD data is easily handled.

Expert systems could enable the total automation of process planning

If the thought processes and decisions of a process planner from the moment he/she looks at a design to the final completion of the plan can be captured in computer logic, the problem of automating the process planning function is virtually solved. Although many research teams are working in this area, most are tackling expert systems for a specific process type, for example turning a shaft. It will therefore be some time before a general-purpose package becomes commercially available.

8 ROBOTICS

By definition a robot is a programmable, multifunctional manipulator designed to move material, parts, tools or specialised devices through variable motions for the performance of a variety of tasks (Robot Institute of America). The important word in this definition is 'programmable'. The functions of a robot are controlled by a programmable microprocessor-based system which can be linked into a computerised work environment. Not only can it process work automatically but it can also react to a changing work environment, receiving and transmitting data relating to its function. Robots therefore are an essential part of any computer integrated manufacturing system.

Robots are an essential part of CIM

In the past, robots have been mainly used in stand-alone activities. In the early days, they performed hot, heavy and hazardous jobs previously performed by humans. A robot once programmed can work at the same rate continuously throughout the day without losing quality. It can work 2–10 times faster than a human and produce fewer faulty parts.

Displacement of people is not the main reason behind robotics

Today, however, the emphasis is changing. Improved productivity and quality, rather than the displacement of people, is the driving force behind the introduction of robots in the workplace. The robot is no longer a stand-alone item, either in its hardware or software environment. In fact, the robot only accounts for about one third of the cost of most robot systems currently being installed. Peripheral items such as work-handling equipment, grippers and electronic interfaces account for the major part of the cost.

ROBOT TYPES

Many types of robot exist, and the range of applications for which they are used is rapidly growing. All models have common features: most are essentially a mechanical arm, fixed to the floor, wall, ceiling or another machine; at the end of the arm is an *end-effector* which may be some kind of gripper, or a tool such as a welding gun or paint spray unit. Movement of the arm may be powered by a number of means: hydraulic, electric or pneumatic *actuators*. Control of the movement is effected by a microprocessor-based controller which senses the position of the arm by monitoring feedback devices on each joint.

Robots can be programmed in different ways. When using the teach method of programming, as opposed to off-line programming, an operator will move the robot arm physically through the desired sequence or alternatively guide it through the motions by remote control. It is possible to program some sophisticated robots directly by telling them to move given distances in given directions.

The essential role of the robot arm is to move a gripper or tool to given orientations at a given set of points. For this

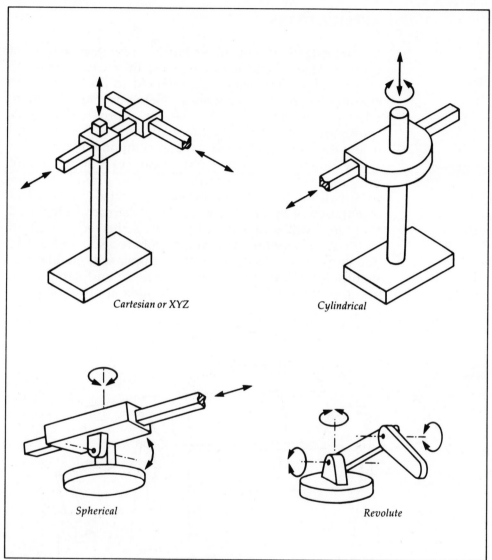

Cartesian or XYZ

Cylindrical

Spherical

Revolute

Fig. 25   The different designs of robot arm

purpose the arm is required to have six articulations, or degrees of freedom: three translational and three rotational. There are four basic designs of robots arms as shown in Fig. 25.

APPLICATIONS

In the past, the use of robots has been limited because of their complete lack of intelligence; they would only do exactly what they were programmed to. This is slowly changing as sensing systems are being developed which, when fitted to the robot system, provide it with information about its environment, such as position and orientation of pallet, product, and components.

Sensing is an essential part of robotics

A vision system, incorporating camera and processing unit, can be linked up with a robot to provide it with information about, for example, the precise position and orientation of a part. This technology is already being used (Fig. 26). Still under development is the technology of tactile sensing which would give robots a sense of touch. Tactile sensors fitted to a robot gripper will enable it to manipulate parts where force, pressure and compliance are important, for example in assembly applications.

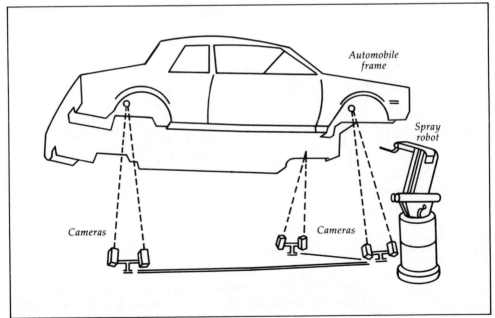

Fig. 26   Stereo location of a car body using cameras to direct a robot in seam sealant spraying

Robots have seldom been used to date in the area of assembly, but it is this application where the greatest growth is expected as the cost of sensing systems drops and as they become less complex to apply. The assembly of printed circuit boards, for example, is expected to be revolutionised by the use of robots.

pcb assembly will be revolutionised by the use of robots

The three areas where robots have been traditionally used are handling, spray painting and welding (Fig. 27). Handling applications are varied, ranging from the loading/unloading of machine tools, presses, diecasting machines, plastic moulding machines, and heat treatment equipment to deburring and fettling operations.

In spray painting, the end-effector is a spray gun which applies a series of thin coats of fast-drying paint. The spray gun must be kept moving and held at a constant distance from the object being sprayed – a job which is ideally suited to a robot, and one to which it has been widely applied particularly in the automotive industry.

The automotive industry has also seen the widespread use of robots for spot welding (sometimes known as resistance welding). Here the end-effector is an electrode

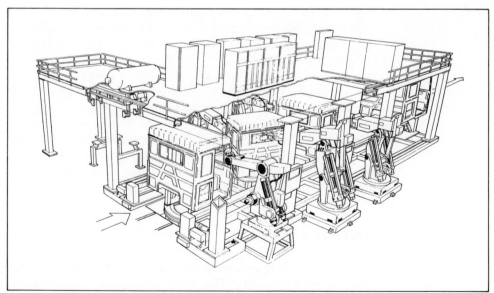

Fig. 27   A four-station robot welding cell

assembly incorporating two electrodes which can hinge together with the sheet metal to be welded between them. The welding process itself is very short so it is vital to move fast between welds to keep non-productive time to a minimum – something a robot is very good at.

TASK SELECTION

Choose robot applications carefully – some tasks are more suitable than others

A robot is not the answer to every production problem. Although any job that can be done by the human hand can be mechanised, some are more suitable for robotic application than others. Firstly, the speed at which a robot works allows a greater throughput of work than is possible by hand. Consideration must therefore be given to whether there is sufficient work available to keep the robot busy. A robot represents a large investment and the aim should be to keep it running almost 24 hours a day (time must be allowed for maintenance and trials). It is possible that a number of different jobs of a similar nature can be assigned to the robot to fill its time.

Another question is work handling. To match the speed at which the robot works, parts entering and leaving the robot workplace must do so at the same rate, and the handling system needs, therefore, to be of a more effective nature.

Justification on reduced manning alone is unlikely

No application of robots should be ventured unless it can be proved to be cost-effective for the company. It is unlikely that a robot application can be justified only on the basis of reduced manpower. The major savings companies should look for when implementing robots are in product improvements. The areas where this can be achieved are in quality, cost and cycle time, all of which are, of course, interlinked.

JUSTIFICATION

Reduction in manpower is, in fact, the easiest cost reduction to quantify. The other savings are indirect and difficult to assign values to. They include:

- Product/process quality improvement and reduced scrap
- Reduced lead-time, controlled lead-time
- Process adaptability
- Reduced inventory
- Reduced cycle time per part

Justification on a single project alone is unrealistic

Most companies insist that return-on-investment (ROI) principles are followed, the payback period required usually being two or three years, but this is a most unrealistic approach. Another misconception of most finance directors is that the introduction of robots can be justified on a specific project. Unless a long run product is being manufactured, a robot as a result of its flexibility, can always be used in a number of different applications during its lifetime. Consequently the cost should be spread over a number of projects. Finance departments often find this hard to accept; and production engineers find it hard to anticipate the varied projects a robot might be employed in during its lifetime. The result is that investment in robotics is most often speculative and it must be accepted as such.

Investment in robotics is often speculative

As stated earlier, the cost of the robot is only a small part of the cost of the system in which it operates. Roughly, the robot purchase price is 30–40% of the cost of the installed robotic system, with the peripheral equipment costing 30–50%. The additional costs of 30–40% are for engineering, installation and training.

Firstly, a robot bought 'off-the-shelf' has to be equipped with grippers that enable it to perform the required task. Each set of grippers is likely to be unique and this will incur engineering costs. Building the handling system which transports the parts to and from the robot will introduce further engineering costs. Although the use of a robot does not necessarily mean that all existing equipment used with a manual system has to be scrapped, it is likely to need serious modification.

The other major cost incurred when implementing a robot system is the cost of programming. It is worthwhile spending some time on preparing programs for a robotic installation since they will be used repeatedly, and any time saving that can be achieved by clever programming will eventually be significant.

PART DESIGN

Product design should proceed concurrently with robotic systems design

Before performing detailed design of a robot system, it is worthwhile taking a detailed look at the design of the parts to be handled in it. A slight modification of a part may for example reduce the complexity of the part handling system considerably. It is likely also that by using a robot greater consistency in the finished product is possible and advantage should be taken of this. One problem which can arise is that parts used in the form of rough castings may come with such a degree of variation that they jam up the handling system. This should be taken into consideration in the design process. In some cases it may be necessary to tighten up the design specification with suppliers.

There is no one solution to an application of a robotic system. Both the hardware and software present a user with a wide selection of choices as regards the robotic element, not to mention the peripheral equipment which should be specifically designed for the application. In every case, the final design of a system is likely to be a compromise between changeover time and flexibility.

Robot systems essentially fall into two types: high volume/low mix and low volume/high mix. In both cases, it is desirable to keep changeover times to a minimum and the aim is therefore to design the system for optimum flexibility. However, the more flexible the system, the more costly it will be; hence a compromise solution results.

SYSTEM SIMULATION

Use both robot simulation and systems simulation before committing to a scheme

Simulation is one way of assessing different solutions for a robot-based application. The standard packages offered by many CAD/CAM and robot vendors are usually graphics-based and dynamic, thus enabling the user to set up a model of the system on a screen and observe its

Fig. 28   Robot simulation setup (courtesy of Calma Co.)

performance (Fig. 28). Firstly, he can check that none of the moving parts collide or move dangerously close during operation. Secondly, the simulation can indicate how good a performance the robot system is capable of, whether productivity is adequate and whether bottlenecks are occurring in the handling system.

With simulation it is relatively easy to see how different configurations affect the system's performance. The user may wish to experiment by changing the number of robots involved in a system, the type of robots, the distribution of tasks between robots and so on, in an effort to achieve the best compromise on cost, changeover times and flexibility.

SYSTEM CONTROL

Robotic systems should be data driven and software controlled

Different robot applications will have different control requirements. Every robot is supplied with its own control unit which, when forming part of an automated system, will need to exchange information with other equipment. The control unit has therefore to be equipped with input/output ports. In the case of a welding or painting robot, the control unit has to communicate with the handling system presenting the work to it, so that the robot knows when to start the operation and the handling system knows when to deliver the next job. For a robot loading/unloading machine tools/processing equipment, the controllers of both pieces of equipment also need to exchange start/stop information.

One controller in the cell must act as the master

In all these applications, one controller has to be the 'master' of the cell, and it is usually the robot's controller. Thus, in a machine-loading cell, the robot is responsible for moving parts between stations, and it also starts up the various processing equipment, energises functions such as opening machine doors and operating conveyors, counts cycles for tool change purposes and shuts down the cell on receipt of certain warning signals.

This arrangement can really only be satisfactory for quite simple cell configurations where little flexibility is required. Progress announcements are made almost daily for stand-alone robot and programmable machine controllers based on microprocessors which were previously limited by memory capacity, computational power, operator system sophistication and any inability to deal with sensor integration, adaptive control and error recovery requirements in time.

The increased management control demanded by the total CIM hierarchy concept dictates that the robot cell has a supervisory computer to which the robot(s) is linked.

The tasks of the supervisory computer are:

- Material transport control
- Cell sequencing
- Automatic changing of CNC part programs
- Messages to and from operator
- Machine status reporting
- Tool management
- Cell diagnostics
- Quality information

The cell supervisory computer can then link into the factory communications network.

9 AUTOMATED MATERIAL HANDLING

A material-handling system generally incorporates a storage function and a transport function; the design of both is greatly influenced by the volume, variety and turnaround of inventory in a plant and this is particularly so in the CIM environment.

Few companies treat material handling with sufficient importance

The importance of material handling as a part of the manufacturing organisation cannot be overestimated, although there are few companies that give it the attention it deserves. It has been calculated that products spend up to 95% of their time in the factory in moving and waiting activities. Of the remaining 5%, 60% is calculated to be lost in loading/unloading activities, which leaves a total of 98% spent in material-flow activities. The scope for improvement is therefore tremendous.

One of the principle aims of CIM is to maximise asset utilisation; a critical objective is therefore to minimise inventory – raw material, work-in-progress and finished product inventory. There are idealists among those following just-in-time concepts who claim that all inventory should be eliminated; they say that the raw material should be delivered during, say, a set four-hour period, that it should start on the production cycle immediately, and continue from one workstation to another without stopping. Then, as soon as the product is finished it should be dispatched without delay. In such an operation no storage functions are necessary, just a super-efficient transport system.

JUST-IN-TIME

Until CIM is implemented, JIT is generally applicable to repetitive and not small batch production

The just-in-time (JIT) situation described above is extremely idealistic. If a company were ever to achieve such an operation, it is unlikely that it would continue for very long before part of the equipment broke down or some emergency arose causing inventory to accumulate at one or more stages in the manufacturing cycle. It must be accepted that storage will always be an important aspect of manufacturing and that facilities must be able to handle raw materials, work-in-progress and finished products.

The exact location of all raw material, semi-finished part and finished product must be known at all times

In the CIM environment, materials handling becomes more essential since, if CIM is to operate effectively, it is crucial for the exact location of every item of raw material, semi-finished part and finished product to be known. In addition, it is most important that stored items are readily accessible and transported to where they are needed in a controlled manner so that production is never delayed by components not being in the right place at the right time.

The introduction of CIM puts more pressure on a company's material-handling techniques than in traditional manufacturing. With the aim of CIM being to minimise inventory, reduce lead-times and produce to customer order as far as possible, the demands on any

*Store and
distribute
materials received
and finished goods
inventory in a
demand pull
environment for
just-in-time
delivery*

storage system multiply. A company operating along these principles is likely to be handling larger numbers of smaller orders; this will result in more frequent deliveries of smaller volumes of raw material and components. The stores needed to cope with this kind of task need not require excessive volumes of space but they must be highly efficient to perform the large number of transactions expected of them. The amount of material actually being stored may not be high but the movements in and out will be numerous.

AUTOMATION

I N T

As in many sectors, automation has gradually been introduced in material-handling technology, resulting in 'islands of automation'. One example of an island is the automated storage and retrieval system (ASRS) or automated warehouse which many companies have found economically viable. Another is the automated guided vehicle (AGV) system which has been used for delivering parts to different areas of a factory.

Stand-alone automation in material handling has been able to give good returns, partly because traditional methods were poor at maintaining inventory control. Material could be put into a store and taken out without any control and it was a frequent occurrence for material to be completely lost in a store.

Cost savings from the introduction of automation can be expected to come from the following sources:

- *Improved control* – Real-time data about material status should be a feature of every automated handling system. The control of goods in a plant can therefore be highly efficient.
- *Reduced labour* – Labour costs are reduced and productivity enhanced by the use of automated handling techniques such as AGVs, warehouses and robots.
- *Improved use of space* – Automation, particularly in warehouses, has enabled the area occupied to be reduced as a result of the introduction of increasingly

higher stacks and narrower aisles. This is important, as space costs money!

- *Increased throughput* – Automation increases the handling throughput which in turn improves inventory levels and manufacturing productivity.
- *Greater flexibility* – The microprocessors with which automated handling systems are equipped provide great flexibility. With the ability to handle a diverse range of loads, an automated system can give long-term benefits.

HANDLING NEEDS

Plan to achieve material being delivered from the vendor directly to the production line – minimise or eliminate further handling steps

All goods, when not being worked on, must be stored in a controlled manner

Whatever products a company is manufacturing, raw materials enter and finished products leave the plant. Until recently, the stores which companies employed were only intended to handle either raw materials or finished goods (but nothing in between). This meant that work-in-progress piled up untidily on the shopfloors. Not only did it occupy valuable space but it was also liable to be damaged and mislaid.

In the CIM environment, it is crucial that when material is not being worked on it should be deposited in a store. Thus, as soon as a batch has been processed at one workstation it either goes straight to the next workstation or to a store. In this way strict control of the quantity, location and state of material can be maintained.

The traditional method of operator-driven fork-lift trucks for transporting bins of material between stores and workstation and between one workstation and another relies too heavily on manual involvement for the CIM environment. Firstly, it is susceptible to human error – the wrong bin can easily end up in the wrong place without being detected. Secondly, it is difficult to control when the bin will arrive at its destination. In addition, fork-lift trucks are relatively large and cumbersome.

Computer-controlled transport, linked into the manufacturing control network, is now a possible solution offering constant monitoring of vehicle location, ability to meet peak demands, optimisation of routing and predictability of material transfer.

While the traditional store occupied an isolated area remote from any manufacturing operations, the CIM environment requires a more integrated approach. Some stores must be in close proximity to the shopfloor to prevent considerable time being wasted as material is transported back and forth to the shopfloor. By modelling and visual interactive simulation, the optimum store location can be easily determined.

The latest advances in manufacturing technology also present greater demands on storage facilities. Machines already exist which provide enormous flexibility. By changing a gripper on a robot, a chuck on a lathe, or a tool magazine on a machining centre, a workstation can be automatically converted from performing one operation to another. This depends, however, on the right gripper, chuck and tool magazine being in the right place at the right time. The systems which handle the workpieces are also being employed for the peripheral equipment.

TRANSPORT

A plant will rely on one main transport system

While a company is likely to find it necessary to employ several different forms of storage technology, there will probably be one main transport system serving the plant. In fact, this has to be the case if the system is to interface directly with the stores and service a number of different manufacturing areas in the most flexible way possible.

Whatever main transport system is selected, it is vital that it should offer the capability of full computer control and be able to integrate with a host computer. This will then enable the system to give real-time response to the production environment, as well as provide an opportunity to optimise the efficiency of the handling system and to operate unmanned shifts.

The AGV is one solution for this which is beginning to gain tremendous popularity. An AGV is a transporter which is programmable, self-powered and guided by cables buried in the floor (chemical tape or infrared). It can be linked into a host computer which can instruct the vehicle to go to station X, pick up a pallet and take it to station Y. Alternatively, an operator can input such instructions through a terminal.

The AGV is increasingly being adopted in a wide range of applications. For example, in distribution environments, AGVs are used to move large volumes of material over long distances. They can be used, for example, to tow goods from the receiving dock to bulk storage areas. On the shopfloor, AGVs are used to transfer parts and tools between the stores and the workstations. Assembly applications are increasingly using the AGV as a moving platform on which the various assembly operations are performed, sometimes manually, sometimes automatically. Fig. 29 shows an AGV fitted with powered roller conveyors for transporting workpieces.

Fig. 29   AGV fitted with powered roller conveyor

AGVs are a very
flexible solution
One of the great advantages of AGV systems is that they are flexible: firstly, the same circuit and control system can support different designs of vehicle performing different functions; secondly, it is easy to design an AGV system around existing plant facilities, whatever the shape of the shopfloor or location of pillars; thirdly, although some current systems do require a cable guidance, the cost of laying or re-laying the cable is not high. It is therefore practical to change or extend an AGV system layout should product design, market demand, processing technology or factory layout need to be extensively altered.

This flexibility gives the AGV considerable advantage over rail-guided vehicles or conveyors, although both these methods can have uses in the CIM environment. AGVs, however, give much better access to equipment should manual intervention be necessary.

The design of AGVs is now very sophisticated. They are designed in a variety of shapes and sizes to carry loads varying from light printed circuit boards to heavy castings and stacks of sheet metal. Some merely tow their loads around while others carry them and can incorporate complex shuttle/lifting mechanisms to transfer loads automatically on and off load/unload stations.

Precise positioning
of the vehicle is
crucial
The AGV control systems are also highly developed. They have to be able to position the vehicle precisely enough for it to interface with other equipment. Vehicle speed must be highly controllable so that the vehicle can turn corners and also conform to variable throughput rates required in assembly applications.

Each AGV has an on-board battery which inevitably adds extra weight to the vehicle. In some applications it is practical to equip the AGV with a very small battery and program it to monitor battery status and automatically take itself to a recharge station or to a battery replacement station when its power is running out.

STORAGE

The exact storage requirements of a company depend on the type of products it is handling. An electronics firm is likely to employ more carousels and tote stackers than automated warehouses, which will be preferred by the engineering concern.

Distributed storage offers optimum performance

Whatever form of storage is appropriate to a company's activities, it should provide certain features to fit in with the CIM environment: firstly, it should have the flexibility to store variable load types; secondly, it should be able to retrieve specific loads on demand either from instructions via a manual data input terminal or the host computer; and thirdly, a high throughput is essential.

At different stages in the manufacturing cycle, different levels of automation will be possible. Even in the most sophisticated system which can interface directly with an automated transport system, parts kitting is generally still a manual task. A storage system must be able to present a required bin of components to a load/unload station but the job of picking a specific number of components out of the bin has only just been automated in a few advanced electronics assembly factories.

The automated storage and retrieval system (ASRS) is one of the most automated forms of storage to find application so far. Able to operate in an unmanned environment and interface directly with an automated transport system, the ASRS can provide the following savings:

- Labour reductions of 30–40%
- Floor-space reductions of 50–70%
- Related material-handling equipment savings of 60–75%
- System maintenance reductions of 10–50%
- Inventory reductions of 15–20%
- Lost sales reductions of up to 5%
- Inventory shrinkage reductions of 50–75%
- Manufacturing inefficiency reductions of up to 5%

ASRS offers opportunities for unmanned production

The ASRS is being widely used on the shopfloor to operate in conjunction with a flexible machining system, permitting it to run unmanned for extended periods of time. During the manned shifts, operators set up parts with jigs and fixtures on pallets and fill up the ASRS.

AGVs can then shuttle pallets to and from the machine tools, returning finished parts to the ASRS, and allowing the FMS to operate in a fully unmanned mode.

While some FMS installations use an ASRS to store pallets of spare tooling as well as workpieces, other forms of automated tool stores are possible. In some cases a specially designed ASRS devoted to tools may be the best solution for handling the very large varieties often needed in an FMS. The FMS shown in Fig. 30 comprises a tool store and an ASRS, both serviced by an AGV system.

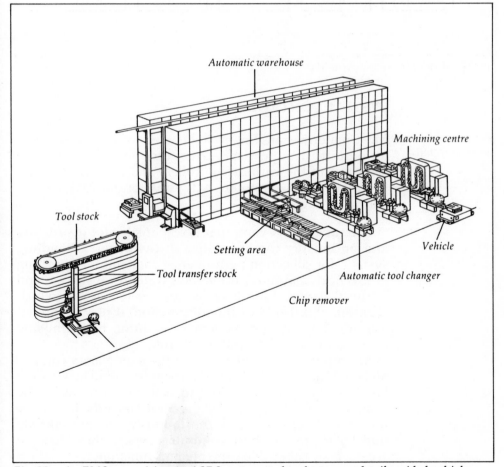

Fig. 30   An FMS comprising an ASRS, automated tools store and rail-guided vehicle

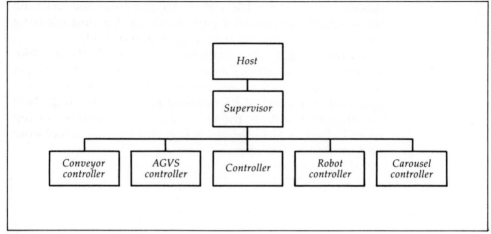

Fig. 31   Typical control hierarchy

SYSTEM CONTROL

Supervisory control is required for complex systems

Most material-handling systems are sufficiently complex to require a supervisory computer to coordinate the activities of the various controllers and to integrate with the main host computer. A typical control hierarchy is shown in Fig. 31.

The supervisory computer is responsible for instructing and optimising the operations of the material-handling system, for collecting performance data and for reporting to the host computer. It tells the warehouse storage and retrieval device what load to store or retrieve in which location, and it retains the information. It plays a similar role for the transport system in addition to optimising routes and providing traffic control.

An important responsibility of the supervisory computer is inventory control. All goods entering and leaving the handling system are tracked in real-time, providing the company with very accurate inventory records. Inventory files which are updated continuously are periodically transferred to the host computer, where they may be accessed by other departments and functions.

In an integrated environment, the material-handling

supervisory computer receives its instructions from the production control function as to which loads are needed, where, and in what order of priority. The material-handling system will also interface with the workstation controller so that when it delivers material to a workstation it can initiate manufacturing operations. Interfaces with the purchasing department are also required in order to keep it informed of the status of deliveries.

AUTOMATIC MATERIAL IDENTIFICATION

Material identification is important in case of system failure

While the fully computer integrated manufacturing system will be entirely paperless, with all information being held in real-time on computer, no system is ever 100% reliable. In an automated material-handling system, numerous mechanical and control elements may go wrong and manual assistance must be available at all times. Some form of material identification is therefore needed even in the most sophisticated of installations.

The most popular method of identification is barcode labelling. A barcode label can record all kinds of data, but for inventory control and work-in-prgoress tracking it is most likely to represent a product number, serial number or an alphanumeric description of the item.

The barcode label, attached to the material, container or pallet, accompanies its load through the system. Hand-held code readers provide the operator with a means of determining a load's identification. Similar readers can be linked into the control network so that barcode labels can be automatically read at various stages in the manufacturing cycle, to communicate information to the production management activity.

SIMULATION

As with many aspects of the plant incorporating CIM principles, simulation is a very important tool for designing and operating the material-handling system. It can help not only to specify, for example, the optimum number of AGVs which are required in an application, but also to develop the control rules to maximise system effectiveness once it is installed.

There are so many variables in a material-handling system that any simulation exercise becomes a full-scale project. Techniques are now readily available which cut the cost and time required to create a model and analyse the results. It is currently possible to model and simulate material-handling systems on personal computers visually and interactively. In this way, it is very easy to try out different layouts and running speeds and immediately perform 'what-if' changes and why any bottlenecks occur. At the same time, the simulation should be able to analyse the performance of each model, providing reports on such topics as AGV and ASRS utilisation.

Simulation of system design is important for optimisation

The quality of the simulation can only be as accurate as the data and breadth of variables used. Every effort should therefore be made to ensure that the model reflects real-life situations, to obtain more than just a simplistic solution.

10 COMMUNICATIONS

Communications is the term used for the transfer of information from one person and/or computer system to another. The term covers a wide range of activities, including the exchange of data between a 'dumb' terminal and a mainframe computer in the same industrial plant, as well as between two computers in different factories in different parts of the world. The information being transferred may be, for example, cutting programs for machine tools, invoice or order details, or electronic mail.

With the decreasing cost and increasing capability of microcomputers, more and more machines incorporating computers have been installed for varying tasks in industry. The need for communications has, therefore, now become of vital importance. But this is only a recent development.

The requirement for communications can be appreciated by considering the situation which developed at General Motors in the USA. To improve the manufacturing process, GM installed large numbers of robots, programmable controllers and other programmable devices, and used statistical process control methods to control 'as built' quality. The large number of programmable devices required computer support for programming, documentation and back-up. In addition, video screens were located around the plant to inform employees about the company (including quality and costs), since workforce involvement is an important element for success.

Additional computers and programmable controllers are used to monitor the cost of utilities such as water, gas, pressurised air, steam, and electricity. They are also used in surveillance tasks to warn of fires and other dangers to the facilities. Finally, as would be expected, GM has accounting systems, personnel systems, material and inventory control systems, warranty systems and others, which use large mainframe computers with remote terminals located throughout the various facilities.

The communications system is the backbone of every CIM implementation

This use of computers by GM is one that all manufacturing industry must adopt in the future and it is one that demands the application of a reliable and extensive communications system. It is this communications system which is the backbone of every CIM implementation and GM realised the importance of the problem – different computers and information generators must be able to communicate with each other.

To understand the present need for communications it is worthwhile looking at how the application of computers has developed from the early days, since many companies are a long way off the advanced stage that GM has achieved. In the 1950s when computers were first used, they were large, expensive and complicated machines which required special isolated rooms, and could only be operated by experts. The users of these machines brought their jobs to the machine to be run in batch. They could not interact directly with their programs while they were running on the computers.

The first interactive access to computers came about

with time-sharing in the 1960s. 'Dumb' terminals were connected to the computer, allowing many people to use the computer at the same time, giving immediate interactive feedback.

More people found more uses for computers as they became more user-friendly resulting in even greater numbers of applications. As demand for the services of these large central computers increased, they had to be upgraded more frequently to provide additional processing power. This was not always practical or possible.

The use of versatile and inexpensive minicomputers became firmly established during the 1970s as users demanded computer power close to the area where the work was being performed. Also for a growing range of applications, users were sharing files, programs, storage and peripheral devices. They needed to exchange data across departments as well as over long distances. A new and greater need arose for communications between one computer and another. These data exchanges required more speed and capacity than the connection of dumb terminals to computers did.

An effective communications system links all the computerised equipment

The power of the central computer began to be distributed among a number of minicomputers. In the 1980s, the advantages of distributing and networking computer systems for office, factory and laboratory environments are being recognised. It is now commonplace for systems to be located at the application site for data processing, database management, process control, typesetting, word-processing and electronic mail, and for these to communicate over limited and long distances via networks based primarily on long-haul communications technology.

TYPES OF COMMUNICATION

The applications which most commonly rely on data communications are:

- *Data entry and collection* – the storage of files for information such as sales data, pay-roll information, inventory control, billing data, and so on.

- *Conversational time-sharing* – for general problem solving, engineering decisions and simple text editing; the user can create files and can alter them and retrieve them at a later date.

- *Remote job entry/batch processing* – a batch of programs are entered at a remote terminal and sent to the host for processing; there is no urgency attached to the job and the host sends back the results of the transactions to a terminal or printer some minutes or even hours after input.

- *Information retrieval* – to obtain answers to transactions immediately, i.e. within seconds.

- *Real-time data monitoring* – when time is of the utmost importance, such as in the monitoring of process technologies.

- *Interprocessor data exchange* – communications between central processing units so that one can take over the moment the other fails.

Communication is needed at several levels in a manufacturing enterprise

It is realised that communication is needed at several levels in a manufacturing enterprise. There are essentially a minimum of four levels of control (see Fig.5 in Chapter One):

- *Companies* – different factories which may be on different sites within a company should be linked.
- *Factories* – the minicomputers in each department should be linked so that they can exchange information.
- *Departments* – a minicomputer in every department should coordinate all the activities of all the computer-controlled devices in that department.
- *Islands* – essentially any item of computer-controlled equipment (machine tool, robot, CAD terminal or word-processor, tape preparation unit, etc.).

It is at the lower levels where the problem of communications is greatest, although distances are shorter; here, more varied devices need to exchange more information. It is also at the local level that databases, programs, files, and the other information assets of a

Different CIM functions have different communications requirements – but all require that correct data arrives at the correct destination at the correct time

company are created, maintained and refined. When communications is accomplished effectively, the whole manufacturing operation becomes greater than the sum of the parts.

Establishing an effective communications network is, however, more difficult than it may at first appear. It is important that messages being originated almost simultaneously receive the correct priority and that accurate data arrives at the desired final destination.

LOCAL AREA NETWORKS

Communication within a plant is effected by a local area network (LAN)

For communications within a single plant, a local area network (LAN) is the commonly used term. The technology of LANs is currently the subject of a great deal of development because of their vital importance in computer integrated manufacturing, and the commercial opportunities have been recognised by the vendors.

Any implementation of a LAN has many different solutions. Therefore, the design of a LAN must be given careful consideration. Depending on the application, speed of communication, for example, may be of higher priority than reliability. Whatever the application, though, some design rules are advisable:

- *Compatibility* – The network should be able to support a wide variety of devices (or nodes as they are often called in this context) all of which need not necessarily be supplied by the same vendor. The LAN should also be able to connect with other networks, for example, so that the communications within different factories of a company can link with one another.

- *Expandability* – Networks should be easily expanded or reconfigured at low cost. This will enable future developments in a company's manufacturing technology to take advantage of existing communications facilities.

- *Reliability* – The entire manufacturing process depends on the smooth running of the communications network.

If this breaks down or develops faults, the entire production operation could come to a halt or, even worse, become totally messed up. Every element in the network, including the node and network interfaces and controllers, and transmission media, should be designed so that failure of a component or node will disable only that unit and not disrupt the rest of the network.

A network architecture takes the form of a seven-layer model

Like the operating system of a computer, a network architecture takes the form of a seven-layer model. Each layer in the hierarchy incorporates a number of modules each of which performs a defined function. The architecture specifies the functions of the modules and the relationships between them.

This structure was defined by the International Standards Organisation in 1978 and has now been widely accepted. It is known as the ISO model for Open Systems Interconnection (OSI) and defines a set of rules of how participating network nodes must interact in order to communicate and exchange information. The architecture defines two kinds of relationships between functional modules:

● *Interfaces* – relationships between different modules that are usually operating within a network node; typically, a module in one layer will interface with a module in the layer below it to receive a service.

● *Protocols* – relationships between equivalent modules, usually in different nodes; protocols define message formats and the rules for message exchange.

Each of the seven layers in the ISO model has certain functions (Fig. 32):

● *Physical link layer* – This layer defines the electrical and mechanical aspects of interfacing to a physical medium for transmitting data, as well as setting up, maintaining, and disconnecting physical links. When implemented, this layer includes the software device driver for each communications device plus the hardware itself (interface devices, modems and communications lines).

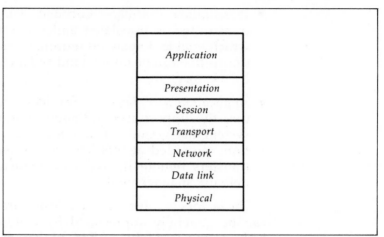

Fig. 32   The ISO seven-layer model for 'open systems interconnection'

- *Data link layer* – This layer establishes an error-free communications path beween network nodes over the physical channel, frames messages for transmission, checks the integrity of received messages, manages access to and use of the channel, and ensures proper sequence of transmitted data.

- *Network control layer* – This layer addresses messages, sets up the path between communication nodes, routes messages across intervening nodes to their destination, and controls the flow of messages between nodes.

- *Transport layer* – This layer provides end-to-end control of a communication session once the path has been established, allowing processes to exchange data reliably and sequentially, irrespective of which systems are communicating or their location in the network.

- *Session control* – This layer establishes and controls system-dependent aspects of communications sessions between specific nodes in the network and bridges the gap between the services provided by the transport layer and the logical functions running under the operating system in a participating node.

- *Presentation control* – Encoded data that has been transmitted is translated and converted into formats which enable display on terminal screens and printers (forms that can be understood and directly manipulated by users).

- *Application/user layer* – Services are provided that directly support user and application tasks and overall system management. Examples of services and applications provided at this level are resource sharing, file transfers, remote file access, database management, and network management.

The factors involved in LAN design are numerous. They can be generally appreciated by looking at one of the hitherto most widely applied – Ethernet.

ETHERNET

Ethernet was developed by Digital Equipment, Xerox and Intel in 1980. It is designed for the high-speed exchange of data between information processing equipment within a moderately sized geographical area and conforms to the seven-layer model.

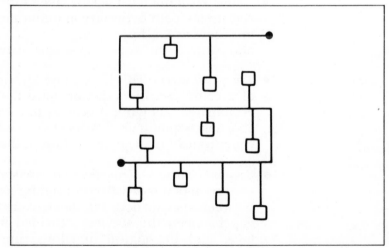

Fig. 33   Ethernet topology – a bus in the shape of a branching tree

The Ethernet network topology is a bus in the shape of a branching tree as shown in Fig. 33. The nodes in the bus topology share a single physical channel via cable taps or connectors. Messages placed on the bus are broadcast out to all nodes, which must be able to recognise their own address in order to receive transmissions.

The medium used by Ethernet is shielded coaxial cable, providing a data rate of 10 million bits per second. It is a baseband LAN which means that signals are sent at their original frequency and only one signal can be transmitted along the medium at any one time.

To ensure that all the nodes do not try to send messages down the common channel simultaneously, Ethernet uses a technique known as 'carrier sense multiple access with collision detect' (CSMA/CD). With this method, nodes can sense whether the channel is busy and can send messages as soon as they sense the channel is free, and should several nodes transmit simultaneously, they have the ability to detect collisions. If a collision is detected, the node waits a short time and then tries again. This method of giving a node access to the transmission medium offers no guarantee of the maximum wait time, although a statistical guarantee can be calculated. This can be a disadvantage in some applications.

Timing of message transmission is critical in the shopfloor environment

The future of Ethernet is more likely to be concentrated in the office environment than on the shopfloor because of its inability to guarantee message transmission time. For the shopfloor environment, timing is critical and for this application, a development spearheaded by General Motors in the USA is becoming a standard approach.

MAP

MAP is a protocol for data communications

In order to achieve a low-cost multi-vendor data communications capability, GM has over the last few years developed a 'manufacturing automation protocol' (MAP) which specifies a LAN based on the seven-layer ISO model especially for shopfloor automation. The demonstration development was initially undertaken in collaboration with Allen-Bradley, Concord Data Systems,

IBM, Hewlett-Packard, Motorola, Digital Equipment and Gould Electronics, all major equipment manufacturers in the USA. Any vendor wishing to supply GM in future must develop equipment which conforms to MAP, lest GM will not consider them. Because of this statement and because of the support GM has received from both vendors and users of automation equipment worldwide, the MAP protocol is expected to become a standard for shopfloor LANs. These LANs will be implemented with gateways to enable them to communicate with the office networks such as TOP (technical office protocol), the standard spearheaded by Boeing, and using a baseband network.

Like Ethernet the MAP network has a bus branching tree topology and uses coaxial cable as the transmission medium. Unlike Ethernet, however, it is a broadband token passing network.

Broadband networks use frequency division multiplexing (FDM) to divide a single physical channel into a number of smaller independent frequency channels, which means that several separate networks are able to operate simultaneously on one physical network channel. They can handle video data and voice communication for a spectrum of applications from closed-circuit TV to NC machine feedback.

Token passing gives the MAP network the guaranteed message transmission time which Ethernet lacks. Instead of CSMA/CD, it allows each node on the network in turn to send a message, thus eliminating the possibility of collisions. A token, which is a special bit pattern or packet, circulates from node to node in a set order. When a node possesses the token it can send a message. When it has finished sending its message, it passes the token onto the next node. In this way, message transmission times are deterministic.

Ensure all new computer-controlled equipment is MAP-compatible

MAP development is still continuing; it is expected to be complete by 1989. Most major vendors and users have already announced their commitment to support this defacto standard. For the user, the situation is not easy but it is prudent to ensure that all new equipment is MAP-compatible. This should enable a wide variety of devices supplied from any of the vendors to communicate on the same network.

DATABASES

As essential to the communications system as the network, is the creation and management of databases. In the past, there has been a preference for one central database to serve an entire company because this would ensure that all data was in one place and would not need to be duplicated. This, however, quickly becomes too large a unit to manage cost effectively to provide a fast and flexible service for the large number of people who would require its use in a CIM environment.

The preferred solution is to have many smaller interfacing databases. This will mean that some data is duplicated but sets of information will be available to the required end-users.

A number of small databases is preferable to a single central one – but data which resides in more than one database must be kept up-to-date in all

A number of factors must be considered when setting up databases. It is vitally important that when data in one database is changed, it is also changed in any other database in which it resides. It is also crucial that access to data is clearly defined. There will be some data that can only be altered by specific personnel, although many others should be allowed access to that data for information purposes only. There will be other data of a high security nature to which almost zero access must be available. Data management, control and security are vitally important aspects of communications which must be addressed within the system specification and design.

11 A STRATEGY FOR CIM – BUSINESS AND PERSONNEL

The very nature of strategic weapons has changed with time. Hundreds of years ago the strength of a nation was determined by the size of its population. In the industrial revolution and the first half of this century, the determining factor was capital assets. Today, it is information and efficient integration of information technology into application systems which is of greatest economic relevance. The strategy for the automated flow of information is, therefore, of vital significance.

Previous chapters in this book describe the CIM technology companies should be aiming for in order to achieve an efficient operation. This final chapter outlines how companies should approach the task of defining a strategy for CIM that suits their own particular needs. CIM requires a very individual approach – there is no one solution, not even for two companies working in the same industry.

With CIM, a company is moulding a future for itself

Starting out on a CIM project is unlike introducing any other new technology into a company. The main difference is that CIM is so all-embracing and wide-reaching. The strategy that a company formulates to introduce it will involve every single department of the organisation. Implementation of the strategy will take place over some period of years, and once complete will set the company on a certain predefined path. Thus, when a company starts out with CIM, it must realise that it is moulding a future for itself.

Establish the highest level corporate strategy first

The route to developing a CIM strategy is largely based on the standard five-step approach taught to industrial engineers: (1) define the problem; (2) collect the data; (3) analyse the data; (4) determine the solution; and (5) implement the solution. For instance, to develop the corporate macro-strategy, the problem may be to determine the corporate identity or ambitions. For the individual business strategy, the problem could be the methods of competition, for the manufacturing unit strategy, it could be the methods of manufacture, and for the CIM strategy, it could be the role of specific computer systems. There are obviously many problems to be identified but until the highest level corporate strategy is formulated and any conflicting 'musts' resolved, the other lower level strategies cannot be optimised.

Fig. 34 shows the general approach to developing CIM planning. In general, most CIM consultants take the four-step approach:

(1) Identify the opportunity or, negatively, the problem
(2) Collect data and assess the situation
(3) Having agreed the solution, refine the planning and develop the specifications for the necessary systems equipment and organisations
(4) Implement – this is performed against the background of the current 'as-is' (existing) situation and determines the 'to-be' (required) situation for three to five years time

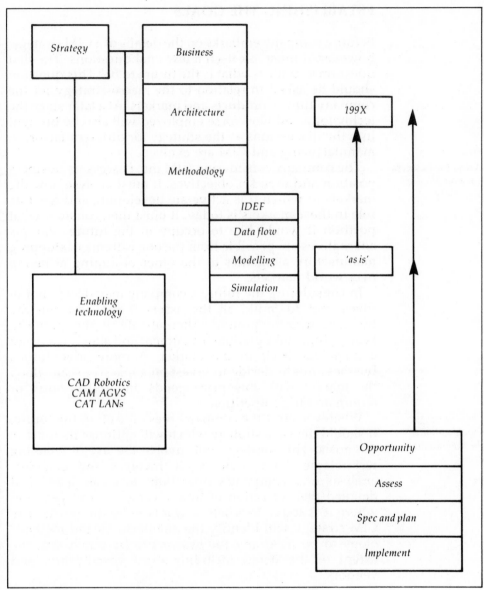

Fig. 34   The general approach to developing CIM planning

The time to complete each of these steps varies from a few weeks, in the first case, to many years for the fourth.

ESTABLISHING THE GOALS

Before a company embarks on the details of a CIM strategy, however, it must ask itself a few vital questions. The first question to settle is what is the future to be? This question should be asked in relation to the macro-strategy for the company future, products and markets. At a later stage the technology and workforce employed will also be brought into the discussions as the strategy for different factories, manufacturing and CIM are evolved.

Assess the business position, then set objectives

The company considering CIM must assess its business position and then set objectives. It must analyse how the markets in which it is active are developing and what its role in these markets is today. It must then consider what position it would like to occupy in the future. Various alternatives are possible from the one extreme of stopping manufacture altogether to the other of aiming at taking over world markets.

In considering the future a company may decide not to invest but to 'milk' all the assets it has and run the business into the ground. Alternatively, it may prefer to keep going, make gradual investment and slowly carve out a niche for itself in the market. A more adventurous business might decide to invest on a massive scale, flood the market with low price goods and out-perform or eliminate the competition.

The strategy should optimise the route to the goals

Whatever view the company sees of itself in the future, it should devise a strategy which will optimise its route to the goals. This strategy will involve the use of enabling technologies and systems. It involves the complete analysis of a company's operations to create a full and detailed plan of action of how a company can get from where it is today to where it wants to be tomorrow. The CIM strategy will identify the architectures and methodology to be used and the systems to be purchased, the extent of the detail including exact specification and vendors.

BUSINESS OBJECTIVES

The company that decides to compete has first to set some

- *Product always meets market demands*
- *A product design which meets specifications and facilitates lowest cost and maximum quality production and maintainability*
- *Minimum engineering and production cycle time*
- *Zero defects*
- *Use of the optimal factors of production*
- *Zero time between manufacturing operations*
- *Zero set-up time*
- *Fewest number of manufacturing operations*
- *Zero raw material and finished goods inventory*
- *Minimal management and support organisational structure*

Fig. 35   CAM-I manufacturing objectives

business objectives. These might include some or all of the following:

- Fast response to market demands
- Better product quality
- Reduced cost
- Enhanced performance
- Better asset utilisation
- Shorter development lead-times
- Minimum work-in-progress
- Flexibility

Fig. 35 shows the manufacturing objectives that CAM-I has set, which would enable the above objectives to be achieved.

Before developing a strategy, an organisation must first resolve an order of priority for its business objectives. This will probably depend on a company's overall commercial strategy. It is interesting to observe the differences at both the international and national level. For example, Japan's first objective seems to be world market dominance at any cost, the USA aims at remaining competitive to satisfy its investors' normal commercial ratios, while the UK attempts to survive.

The strategy is binding but robust enough to absorb change

Companies must consider the strategy they construct to be binding in almost all its detail. The only possible leeway can be allowed if unforeseen events occur, such as new technologies being introduced or existing products being withdrawn. It is therefore crucial that the strategy is

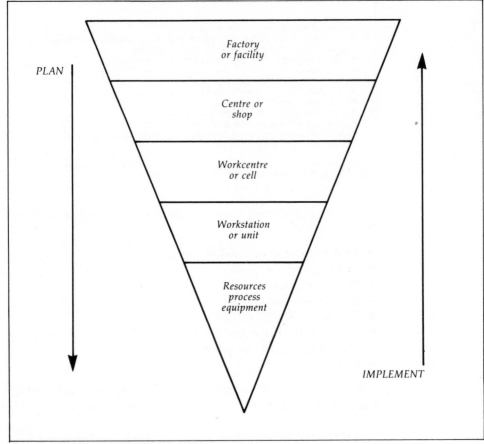

PLAN

Factory
or facility

Centre or
shop

Workcentre
or cell

Workstation
or unit

Resources
process
equipment

IMPLEMENT

Fig. 36 '*Top-down' and 'bottom-up' approaches*

constructed with extreme care and foresight and still be robust enough to absorb changed priorities or new technologies. Conflicting philosophies for business operation must be resolved before the CIM strategy can be finalised.

The strategy must be supported and promoted at the highest company level

CIM strategies are developed with a top-down approach and implemented from the bottom-up, as shown in Fig. 36. They must be supported and promoted at the highest company level. Several aids to establishing this all-important part of a CIM implementation have been created. One is a methodology known as IDEF.

IDEF AS A DECISION AID

IDEF is important as a decision aid

IDEF is a tool developed by the US Air Force to assist companies to analyse and evaluate an organisation and to plan and implement its improvement. It consists of two parts: IDEF 0 and IDEF 1. IDEF 0 is a function modelling tool which analyses the organisation by function using a hierarchical top-down approach. Each function is represented by a rectangular box and performs transformations on physical items and/or information (inputs/outputs) by means of resources (mechanisms) under the influence of some control.

IDEF 1 is an information modelling tool which is complementary to IDEF 0 and employs a bottom-up approach. Data sources, uses and inter-relationships are analysed to determine the documentation required, and define file and database structures which generate minimum duplication.

Using IDEF, anomalies, duplication and inefficiencies can be brought to light

Using these two IDEF tools, a thorough understanding of how an organisation works can be gained. At this stage, the company is likely to be able to identify anomalies, duplication and inefficiencies that have crept into the operation over a period of time. With these examples of bad practice brought into the open, the company is now in a much better position to plan its strategy to reach its goals of the future.

IDEF methodology has been adopted by different companies in different ways. Computer aids for generating the IDEF relevant to a specific company are available but many companies may find that the skills required to do the model generation are already to be found in their existing software development organisations. Software development tools are equally applicable to strategy and operational development.

SIMULATION

Simulation allows alternatives to be considered

In drawing up the detailed strategy, companies should ensure that the final plan is robust enough to withstand change in the boundary conditions, e.g. cash-flow changes, program delays, market conditions, enabling technology. Alternatives must be considered as the strategy is built up. A tool to help assess the best solutions is simulation.

Simulation and modelling can be used in a number of ways, such as a model of the proposed factory layout can be used to optimise the use of space and minimise the movement of goods and personnel. Another application might be to test out the performance of production systems. Does the system give the desired throughput of parts? Should another machining centre or component insertion machine be added? Is there sufficient in-process buffer to allow unmanned operation for a complete shift? These are just some of the 'what-if' questions that simulation should be able to answer. Perhaps, even more important, is the use of simulation to investigate what can happen in hypothetical emergency situations such as failure of a robot for some period of time or the market situation changes and a new product mix is demanded at short lead-times.

A number of simulation packages are available to help companies through this stage. They are almost all interactive dynamic graphic systems which are easy to use by non-computer experts both for creating models and changing the conditions to see what happens.

HUMAN RESOURCES

Almost as important as forming the right strategy for the CIM implementation is the management of human resources. Human motivations have as much influence over the success or failure of the project as finance and technology.

CIM breaks down the barriers between departments

CIM creates new relationships within an organisation, its main effect being to break down the clearly defined divisions between departments. These new relationships come into play both when planning the project as well as when implementing and running it.

Teamwork is essential

Teamwork is essential. Engineering, manufacturing and data processing departments must work together in planning their systems. Each must understand the others' needs and limitations. Only with mutual agreement and understanding will the most effective systems be purchased and implemented, and finally put into successul operation. This will undoubtedly mean compromises on the part of one department or another, therefore a spirit of trust and respect must exist.

Complete workforce support is essential for success

If a major change such as CIM is to have any chance of success, it must have complete workforce support. At the earliest planning stage possible the labour force should be informed and consulted. Not only does the workforce possess valuable knowledge essential to the planning of the new system but it will also be expected to operate the new technology and ultimately has the power to make or break it. The greater the involvement and contribution from the lower levels, the better the chance of a smooth running project.

At the other end of the scale, top management commitment is essential. The workforce must feel that management is fully behind the introduction of CIM and is confident of its success.

A multi-disciplinary team should spearhead the project

Spearheading any CIM project must be a multi-disciplinary team. Not only will this team include representatives from the different departments but it will also feature a mixture of personalities. Those of an adventurous, enthusiastic nature are essential to keep momentum going, while others of a pragmatic and sceptical outlook must also be included. As the CIM project progresses through its different phases, the skill mix and roles of the team members will change. In the early stages when the requirements are being planned and the concept designed, the work is more exploratory and analytical. During detailed design and implementation, it is more technical.

Large organisations may find it worthwhile to set up a specialist team to play a consultancy role in the organisation in planning and implementing CIM projects. This can be valuable for a company located on different sites where the specialist team can be useful in technology transfer.

The sensitive aspect of personnel management in the CIM context is its effect on numbers – labour reduction is almost always a factor. Linked with this is the change in job description. Many of the totally unskilled jobs, such as machine loading, will be eliminated but they will be replaced by jobs of greater responsibility because the machine loading operator in the CIM context is likely to look after a number of machines, use a terminal to receive and process data, and handle minor maintenance problems that occur.

As far as possible, existing staff should be trained to take on the new jobs created

A company should, as far as possible, train its existing staff to take on the new jobs created by the introduction of CIM. This will make the whole project more acceptable to the trade unions but, in addition, staff who have worked for some years in a company know its products and methods – which is highly valuable.

BIBLIOGRAPHY &
FURTHER READING

Auerbach Publishers, Inc. 1985. *CAD/CAM Technology Series: CAD/CAM Management Strategies; Computer-Aided Design, Engineering and Drafting; Automated Material Handling and Storage; Computers in Manufacturing*. Auerbach, Pennsauken, New Jersey.

Ayres, R.U. and Miller, S.M. 1985. *Robotics and Flexible Manufacturing Technologies: Assessment, Impact and Forecast*. Noyes Publications, Park Ridge, New Jersey.

Burbridge, J.L. 1975. *The Introduction of Group Technology*. Heinemann, London.

Butterworth, D.D. and Bailey, J.E. 1982. *Integrated Production Control Systems. Management, Analysis, Design*. John Wiley & Sons, New York.

CAM-I, 1985. Library Catalogue R-85-AMP-01. Computer Aided Manufacturing International, Arlington, Texas.

Charles Stark Draper Laboratory, 1984. *Flexible Manufacturing Systems Handbook*. Noyes Publications, Park Ridge, New Jersey.

Chase, R.B. and Aquilano, N.J. 1981. *Production and Operations Management*, 3rd edition. Richard D. Irwin, Homewood, Illinois.

Childs, J.J. 1982. *Principles of Numerical Control*. Harper & Row, New York.

CIM Review – The Journal of Computer-Integrated Manufacturing Management. Auerbach Publishers, Inc.

Computer Aided Manufacturing International, Inc. 1982. *Process Planning Program*, PR-81-ASPP-01.3, Arlington, Texas.

Digital Equipment Corp. 1982. *Introduction to Local Area Networks*. Digital Equipment Corp., Bedford, Massachusetts.

FMS Magazine. IFS (Publications) Ltd.

Groover, M.P. 1980. *Automation, Production Systems and Computer-Aided Manufacturing*. Prentice-Hall, Englewood Cliffs, New Jersey.

Groover, M.P. and Zimmers, E.-W. 1984. *CAD/CAM: Computer-Aided Design and Manufacturing*. Prentice-Hall, Englewood Cliffs, New Jersey.

Halevi, G. 1980. *The Role of Computers in Manufacturing Processes*. John Wiley & Sons, New York.

Hammond, G. 1986. *AGVS at Work*. IFS (Publications) Ltd.

Harrington, J. 1979. *Computer Integrated Manufacturing*. Krieger.

Hartley, J. 1984. *FMS at Work.* IFS (Publications) Ltd.

Hitomi, K. 1979. *Manufacturing Systems Engineering.* Taylor & Francis, Philadelphia, Pennsylvannia.

Hollingum, J. 1986. *The MAP Report.* IFS (Publications) Ltd.

Hundy, B.B. (Ed.) 1985. *Proceedings of the 3rd European Conference on Automated Manufacturing.* IFS (Publications) Ltd.

Ingersoll Engineers, 1985. *Integrated Manufacture.* IFS (Publications) Ltd.

Ingersoll Engineers, 1982. *The FMS Report.* IFS (Publications) Ltd.

Institution of Production Engineers, 1985. *A Guide to CAPM.* I. Prod. E, London.

International Journal of Advanced Manufacturing Technology. IFS (Publications) Ltd.

Kochan, D. 1986. *CAM – Developments in Computer-Integrated Manufacturing.* Springer-Verlag, Berlin.

Lenz, J.E. (Ed.) 1986. *Proceedings of the 2nd International Conference on Simulation in Manufacturing.* IFS (Publications) Ltd.

Lindholm, R. (Ed.) 1985. *Proceedings of the 4th International Conference on Flexible Manufacturing Systems.* IFS (Publications) Ltd.

Machover, C. and Blauth, R.E. (Eds.) 1980. *The CAD/CAM Handbook.* Computervision Corp., Bedford, Massachusetts.

Martin, J. 1983. *Managing the Database Environment.* Prentice-Hall, Englewood Cliffs, New Jersey.

McGeough, J.A. (Ed.) 1986. *Proceedings of the International Conference on Computer-Aided Production Engineering.* Mechanical Engineering Publications Ltd, Bury St. Edmunds, UK.

Medland, A.J. and Burnett, P. 1986. *CADCAM in Practice: A Manager's Guide to Understanding and Using CADCAM.* Kogan Page, London.

Nazemetz, J.W., Hammer, Jr., W.E. and Sadowski, R.P. (Eds.) 1985. *Computer Integrated Manufacturing Systems: Selected Readings.* Industrial Engineering and Management Press/ Institute of Industrial Engineers, Norcross, Georgia.

Owen, A. 1985. *Strategic Issues in Automated Production : The Challenge of Robotic and Computer Integrated Manufacturing.* Cranfield Press, Bedford, UK.

Plessey Co. *A Guide to Testing Surface Mount pcb Assemblies.* The Computer Aided Test Working Group of the Methods Panel, Plessey Co.

Plessey Co. *A Guide to Testing pcbs containing VLSI.* The Computer Aided Test Working Group of the Methods Panel, Plessey Co.

Pusztai, J. 1983. *Computer Numerical Control.* Prentice-Hall, Englewood Cliffs, New Jersey.

Ranky, P.G. 1986. *Computer Integrated Manufacturing: An Introduction with Case Studies.* Prentice-Hall, Englewood Cliffs, New Jersey.

Ranky, P.G. 1983. *The Design and Operation of FMS.* IFS (Publications) Ltd.

Rembold, U. and Dillmann, R. (Eds.) 1984. *Methods and Tools for Computer Integrated Manufacturing.* Springer-Verlag, New York.

Rembold, U., Blume, C. and Dillmann, R. 1985. *Computer-Integrated Manufacturing Technology and Systems.* Marcel Dekker, New York.

Smith, W.A. (Ed.) 1983. *A Guide to CAD/CAM.* Institution of Production Engineers, London.

Steudel, H.J. 1984. Computer aided process planning: Past, present and future. *Int. J. Production Research*, 22(2):253-256.

Stover, R. 1984. *An Analysis of CAD/CAM Applications with an Introduction to CIM*. Prentice-Hall, Englewood Cliffs, New Jersey.

Yeomans, R.W., Choudry, A. and Ten Hagen, P.J.W. (Eds.) 1985. *Design Rules for a CIM System*. North-Holland, Amsterdam.

GLOSSARY

adaptive control A control method in which control parameters are continuously and automatically adjusted in response to measured process variables to achieve near-optimum performance.

advanced manufacturing technology (AMT) General term to cover the application of various developing techniques such as CAD, FMS, robotics and ATE to industrial manufacturing.

algorithm In CAD/CAM software, a set of well-defined rules or procedures based on mathematical and geometric formulae for solving a problem or accomplishing a given result in a finite number of steps.

analogue transmission Transmission of a continuously variable signal as opposed to a discretely variable signal such as digital data. Examples of analogue signals are voice calls over the telephone network, facsimile transmission, and electrocardiogram information.

application program (or package) A computer program or collection of programs to perform a task or tasks specific to a particular user's need or class of needs.

artificial intelligence (AI) The capability of a computer to perform functions that are normally attributed to human intelligence, such as learning, adapting, recognising, classifying, reasoning, self correction and improvement.

artwork One of the outputs of a CAD system. For example, a photo plot (in pcb design), a photo mask (in IC design), a pen plot, an electrostatic copy, or a positive or negative photographic transparency. Transparencies (either on glass or film) and photo masks are forms of CAD artwork which can be used directly in the manufacture of a product (such as an IC, pcb, or mechanical part).

attribute A non-graphic characteristic of a part, component, or entity under design on a CAD system. Examples include: dimension entities associated with geometry, text with text nodes, and nodal lines with connect nodes. Changing one entity in an association can produce automatic changes by the system in the associated entity; i.e. moving one entity can cause moving or stretching of the other entity.

automated guided vehicle system (AGVS) Vehicles equipped with automatic guidance equipment which follow a prescribed guidepath which interfaces with workstations for automatic or manual loading and unloading. In FMS, guided vehicles generally operate under computer control.

automated handling system System(s) used to automatically move and store parts and raw materials throughout the manufacturing process and to integrate the flow of workpieces and tools with the manufacturing process. In FMS, the automated material handling system operates under computer control.

automated storage and retrieval system (ASRS) A high density rack storage system with rail running vehicles serving the rack structure for automatic loading and unloading. Vehicles interface with AGVS, car-on-track, towline or other conveyor systems for automatic storage and retrieval of loads. In FMS, automated storage systems operate under computer control.

bill of materials (BOM) Manufacturing data referring to parts and materials that list what they are used for, how frequently and how they are structured.

buffer inventory An inventory within a manufacturing process maintained as a protection against the effects of fluctuations of supply and demand, e.g. raw materials, finished stock or work-in-progress.

business information systems The commercial systems within a company such as sales order processing, sales ledger, purchase ledger, payroll, nominal ledger and sales analysis.

CAD (computer aided design) Describes the more demanding and elaborate preparation of complex schematics and blueprints, typically those of industry. In these applications, the operator constructs a highly detailed drawing on-line using a variety of interaction device and programming techniques. Facilities are required for replicating basic figures; achieving exact size and placement of components; making lines of specified length, width, or angle to previously defined lines; satisfying varying geometric and topological constraints amongst components of the drawing; etc. A primary difference between interactive plotting and design draughting lies in the amount of effort the operator contributes, with interactive design draughting requiring far more responsibility for the eventual result. In interactive plotting, the computation is of central importance and the drawing is typically secondary. A second difference is that design drawings tend to have structure, i.e. to be hierarchies of networks or mechanical or electrical components. These components must be transformed and edited. If, in addition to non-trivial layout, the application program involves significant computation of the picture and its components, we speak of the third and most complex category, that of interactive design. In addition to a pictorial datum base, or data structure, that defines where all the picture components fit on the picture and also specifies their geometric characteristics, an application datum base is needed to describe the electrical, mechanical, and other properties of the components in a form suitable for access and manipulation by the analysis program. This datum base must naturally also be editable and accessible by the interactive user.

CAD/CAM   Refers to the integration of computers into the entire design-to-fabrication cycle of a product or plant.

CAE (computer aided engineering) Analysis of a design for basic error-checking or to optimise manufacturability, performance, and economy (for example, by comparing various possible materials or designs). Information drawn from the CAD/CAM design database is used to analyse the functional characteristics of a part, product, or system under design, and to simulate its performance under various conditions. CAE permits the execution of complex circuit loading analyses and simulation during the circuit definition stage. CAE can be used to determine section properties, moments of inertia, shear and bending moments, weight, volume, surface area, and centre of gravity. CAE can precisely determine loads, vibration, noise, and service life early in the design cycle so that components can be optimised to meet those criteria. Perhaps the most powerful CAE technique is finite element modelling.

CAM (computer aided manufacturing) The use of computer and digital technology to generate manufacturing-oriented data. Data drawn from a CAD/CAM database can assist in or control a proportion of all of a manufacturing process, including numerically controlled machines, computer-assisted parts programming, computer-assisted process planning, robotics, and programmable logic controller. CAM can involve production programming, manufacturing engineering, industrial engineering, facilities engineering, and reliability engineering (quality control). CAM techniques can be used to produce process plans for fabricating a complete assembly; to program robots; and to coordinate plant operation.

CIM (computer integrated manufacturing) The concept of a totally automated factory in which all manufacturing processes are integrated and controlled by a CAD/CAM system. CIM enables production planners and schedules, shopfloor foremen, and accountants to use the same database as product designers and engineers.

computer aided process planning (CAPP) An application program that is interactive with CAD/CAM and assists in the development of a process/production plan for manufacturing.

computer numerical control (CNC) A technique in which a machine-tool control uses a minicomputer to store NC instructions generated earlier by CAD/CAM for controlling the machine.

continuous-path control A control scheme whereby the inputs or commands specify every point along a desired path of motion. Continuous-path-control techniques can be divided into three basic categories based on how much information about the path is used in the motor control calculations. The first is the conventional or servo-control approach. This method uses no information about where the path goes in the future. The controller may have a stored representation of the path it is to follow, but for determining the drive signals to the robot's motors all calculations are based on the past and present tracking errors. This is the control design used in most of today's industrial robots and process control systems. The second approach is called preview control, also known as 'feed-forward' control, since it uses some knowledge about how the path changes immediately ahead of the robot's current location, in addition to the past and present tracking error used by the servo controller. The last category of path control is the 'path planning' or 'trajectory calculation' approach. Here the controller has available a complete description of the path the manipulator should follow from one point to another. Using a mathematical-physical model of the arm and its load, it pre-computes an acceleration profile for every joint, predicting the nominal motor signals that should cause the arm to follow the desired path. This approach has been used in some advanced research robots to achieve highly accurate coordinated movements at high speed.

control hierarchy A relationship of control elements whereby the results of higher-level control elements are used to command lower-level elements.

CPU (central processing unit) That part of a computer or programmable controller that controls the interpretation and execution of instructions. In general, the CPU contains the following elements; arithmetic-logic unit (ALU); timing and control; accumulator; scratch pad memory; program counter and address stack; and instruction register. This CPU is sometimes referred to as the processor.

database A comprehensive collection of interrelated information stored on some kind of mass data storage device, usually a disk. Generally consists of information organised into a number of fixed-format record types with logical links between associated records. Typically includes operating system instructions, standard parts libraries, completed designs and documentation, source code, graphic and application programs, as well as current user tasks in progress.

data base management system (DBMS) A package of software programs to organise and control access to information stored in a multiuser system. It gives users a consistent method of entering, retrieving and updating data in the system, and prevents duplication and unauthorised access to stored information.

data processing A procedure for collecting data and producing data in format.

direct numerical control (DNC) A system in which sets of NC machines are connected to a mainframe computer to establish a direct interface between the DNC computer memory and the machine tools. The machine tools are directly controlled by the computer without the use of tape.

distributed processing Refers to a computer system that employs a number of different hardware processors, each designed to perform a different sub-task on behalf of an overall program or process. Ordinarily, each task would be required to queue up for a single processor to perform all its needed operations. But in a distributed processing system, each task queues up for the specific processor required to perform its needs. Since all processors run simultaneously, the queue wait period is often reduced, yielding better overall performance in a multitask environment.

dumb terminal An input/output terminal connected to a computer with no capability itself to run stand-alone programs.

end-effector An actuator, gripper, or driven mechanical device attached to the end of a manipulator by which objects can be grasped or otherwise acted upon.

end-of-arm speed There are various opinions on arm speed, varying depending on the axes about which the arm is moving, its position in the work envelope, and the load being carried. Keeping that in mind, a reasonable, though somewhat simplistic question to ask is: How fast can the gripper get from an arbitrary point *A* to an arbitrary point *B* in the envelope, empty, and how fast can it move back from *B* to *A* fully loaded? The best answer that can be expected is a ballpark figure, unless there is a willingness to accept the answer in the form of a differential equation or something similar.

expert system A computer system with the ability to modify the program or data as a result of computations so that it 'learns by experience' and provides new solutions to developing situations.

family of parts A collection of previously designed parts with similar geometric characteristics (i.e. line, circle, ellipse) but differing in physical measurement (i.e. height, width, length, angle). When the designer preselects the desired parameters, a special CAD program creates the new part automatically, with significant time savings.

finite element modelling (FEM) The creation on the system of a mathematical model representing a mechanical part or physical construction under design. The model, used for input to a finite element analysis (FEA) program, is built by first subdividing the design model into smaller and simpler elements such as rectangles, triangles, bricks, or wedges which are inter-connected. The finite element model is comprised of all its subdivisions or elements, and its attributes (such as material and thickness), as well as its boundary conditions and loads (including mechanical loadings, temperature effects, and materials fatigue).

flexible manufacturing system (FMS) An arrangement of machines (usually NC machining centres with tool changers) interconnected by a transport system. The transporter carries work to the machines on pallets or other interface units so that accurate work-machine registration is rapid and automatic. A central computer controls machines and transport. It may have a variety of parts being processed at any one time.

floor-to-floor time The total time elapsed for picking up a part, loading it into a machine, carrying out operations, and unloading it (back to the floor, bin, pallet, etc.); generally applies to batch production.

gripper A 'hand' of a robot which picks up, holds, and releases the part of the object being handled. Sometimes referred to as a manipulator or compliance.

group technology A coding and classification system used in CAD for combining similar, often-used parts into families. Group technology facilitates the location of an existing part with specified characteristics, and helps to standardise the fabrication of similar parts. Grouping of similar parts in a family allows them to be retrieved, processed, and finally fabricated in an efficient, economical batch mode.

hard automation Use of specialised machines and machine lines to manufacture and assemble products or components. Normally each machine or line is dedicated to one function, such as milling. Hard automation is usually of a continuous manufacturing type and is used for high volume production in distinction to batch production.

high-level language A problem oriented porogramming language using words, symbols, and command statements which closely resemble English language statements. Each statement typically represents a series of computer instructions. Relatively easy to learn and use, a high-level language permits the execution of a number of subroutines through a simple command. Examples are BASIC, FORTRAN, PL-1, PASCAL, and COBOL. A high-level language must be translated or compiled into machine language before it can be understood and processed by a computer.

information technology (IT) The microelectronics based combination of telecommunications and computing.

initial graphics exchange specification (IGES) An interim CAD/CAM database specification until the American National

Standards Institute develops its own specification. IGES attempts to standardise communication of drawing and geometric product information between computer systems.

integrated computer aided manufacturing (ICAM) A programme and development plan (sponsored by the US Air Force in cooperation with the Aerospace Industry) for producing systematically related modules of machining for flexible manufacturing and control. See FMS, CAD/CAM.

interactive graphics system (IGS) or interactive computer graphics (ICG) A CAD/CAM system in which the workstations are used interactively for computer aided design and/or draughting, as well as for CAM, all under full operator control, and possibly also for text-processing, generation of charts and graphs, or computer aided engineering. The designer (operator) can intervene to enter data and direct the course of any program, receiving immediate visual feedback via the CRT. Bilateral communications is provided between the system and the user(s). Often used synonymously with CAD.

island of automation Stand-alone automation products, (robots, CAD/CAM systems, NC machines) without the integration required for a cohesive system.

job shop A manufacturing facility which specialises in one-of-a-kind or limited production (small batch processing) of parts and subassemblies or products.

kanban Japanese for just-in-time scheduling system which dramatically reduces work-in-process inventory by delivery of raw materials, parts and subassemblies to production in small batches as they are needed.

level of automation The degree to which a process has been made automatic. Relevant to the level of automation are questions of automatic failure recovery, the variety of situations that will be automatically handled, and the situation under which manual intervention or action by humans is required.

LSI (large scale integration) Any integrated circuit which has more than 100 equivalent gates manufactured simultaneously on a single slice of semi-conductor material. LSI circuits can range up to several thousand logic elements on a one-tenth square inch silicon chip.

materials requirements planning or manufacturing resources planning (MRP) A variety of computer applications for ordering materials and managing inventories that are increasingly interactive with on-line manufacturing data and financial data. See bill of materials.

mean-time-between-failures (MTBF) The average time that a device will operate before failure.

MRP (materials requirements planning) A time-phased, level by level, netting and batching materials planning system.

MRP II (manufacturing resource planning) A development of MRP which takes into account the planned availability of capacity.

network An arrangement of two or more interconnected computer systems to facilitate the exchange of information in order to perform a specific function. For example, a CAD/CAM system might be connected to a mainframe computer to off-load heavy analytic tasks. Also refers to a piping network in computer aided plant design. 'Local networking' is the communications network internal to a robot. 'Global networking' is the ability to provide communications connections outside of the robot's internal system.

numerical control (NC) A technique of operating machine tools or similar equipment in which motion is developed in response to numerically coded commands. These commands may be generated by a CAD/CAM system on punched tapes or other communications media. Also, the processes involved in generating the data or tapes necessary to guide a machine tool in the manufacture of a part.

part family A set of discrete products that can be produced by the same sequence of machining operations. This term is primarily associated with group technology.

post-processor A software program or procedure which formats graphic or other data processed on the system for some other purpose. For example, a postprocessor might format cutter centreline data into a form which a machine controller can interpret.

printed circuit board (pcb) A baseboard made of insulating materials and an etched copper-foil circuit pattern on which are mounted ICs and other components required to implement one or more electronic functions. Printed circuit boards plug into a rack or subassembly of electronic equipment to provide the brains or logic to control the operation of a computer, or a communications system, instrumentation, or other electronic systems. The name derives from the fact that the circuitry is connected not by wires but by copper-foil lines, paths, or traces actually etched onto the board surface. CAD/CAM is used extensively in pcb design, testing, and manufacture.

process planning Specifying the sequence of production steps, from start to finish, and describing the state of the workpiece at each workstation. Recently CAM capabilities have been applied to the task of preparing process plans for the fabrication or assembly of parts.

process simulation A program utilising a mathematical model created on the system to try out numerous process design iterations with real-time visual and numerical feedback. Designers can see on the CRT what is taking place at every stage in the manufacturing process. They can therefore optimise a process and correct problems that could affect the actual manufacturing process downstream.

programmable logic controller (PLC) A stored program device intended to replace relay logic used in sequencing, timing, and counting of discrete events. Instead of physical wiring relay, pushbuttons, limit switches, etc., a PLC is programmed to test the state of input lines, to set output lines in accordance with input state, or to branch to another set of tests. The instruction sets of these machines generally exclude all arithmetic and Boolean operators, but do include vital decision instructions such as skip, transfer unconditional, transfer conditional and even transfer and link.

quality assurance (QA) Denotes both quality control and quality engineering.

quality control (QC) The establishing and maintaining of specified quality standards for products.

quality engineering (QE) The establishment and execution of tests to measure product quality and adherence to acceptance criteria.

robot A reprogrammable multifunctional manipulator designed to move material, parts, tools or specialised devices, through variable programmed motions for the performance of a variety of tasks.

robot motions Four types of work motions: (1) anthropometric motion – motions of a robot as in a shoulder, elbow and a wrist, developing a modified spherical work envelope; (2) cylindrical motion – the motion of a robot's arm when mounted on cylindrical axis; (3) polar motion – the motions of a robot by two axes of rotation which create a modified spherical work envelope; (4) rectilinear motion – motions of a robot in three dimensions along straight lines (slides or channels).

SCARA robot (selective compliance assembly robot arm) A low-cost, high-speed assembly robot moving almost entirely on a horizontal plane.

sensory control Control of robot based on sensor readings. Several types can be employed: sensors used in threshold tests to terminate robot activity or to branch to another activity; sensors used in a continuous way to guide or direct changes in robot motions; sensors used to monitor robot progress and to check for task completion or unsafe conditions; and sensors used to retrospectively update robot motion plans prior to the next cycle.

simulation A CAD/CAM computer program that simulates the effect of structural, thermal, or kinematic conditions on a part under design. Simulation programs can also be used to exercise the electrical properties of a circuit. Typically, the system model is exercised and refined through a series of simulation steps until a detailed, optimum configuration is reached. The model is displayed on a CRT and continually updated to simulate dynamic motion or distortion under load or stress conditions. A great variety of materials, design configurations, and alternatives can be tried out without committing any physical resources.

time-sharing The use of a common CPU memory and processing capabilities by two or more CAD/CAM terminals to execute different tasks simultaneously.

turnkey A CAD/CAM or robotic system for which the supplier/vendor assumes total responsibility for building, installing, and testing both hardware and software, and the training of user personnel. Also loosely, a system which comes equipped with all the hardware and software required to do a specific application or applications. Usually implies a commitment by the vendor to make the system work, and to provide preventive and remedial maintenance of both hardware and software. Sometimes used interchangeable with standalone, although standalone applies more to a systems architecture than to terms of purchase.

VAL A proprietary robot-oriented language developed by Unimation Inc.

vertical integration A method of manufacturing a product, such as a CAD/CAM, system, whereby all major modules and components are fabricated in-house under uniform company quality control and fully supported by the system vendor.

vision The ability to scan and/or look at a given piece or part, usually for welding (but sometimes for orientation), and determine where said part is in relation to where it should be. Then that information is digested and the logic of the system tells the robot what to do.

VLSI (circuit) (very large scale integrated circuit) High-density ICs characterised by relatively large size (perhaps 1/4 inch on a side) and high complexity (10,000 to 100,000 gates). VLSI design, because of its complexity, makes CAD a virtual necessity.

work-in-progress Products in various stages of completion throughout the production cycle, including raw material that has been released for initial processing and finished products awaiting final inspection and acceptance for shipment to a customer.

workstation The work area and equipment used for CAD/CAM operations. It is where the user interacts (communicates) with the computer. Frequently consists of a CRT display and an input device as well as, possibly, a digitiser and a hard copy device. In a distributed processing system, a workstation would have local processing and mass storage capabilities. Also called a terminal or design terminal.